Artistic Interpretations of Science

Exploring Microscopic Life and Quantum Wonders Through an Artistic Lens

Elysia Heartwood

Contents

Contents

1 Introduction to Art & Science

The intertwined history of art and science stretches back centuries, a dance between observation and interpretation. Consider Leonardo da Vinci, a master of both artistic expression and scientific inquiry. His anatomical drawings, meticulous and beautiful, were driven by a desire to understand the human machine, not just to represent it aesthetically. This pursuit of knowledge, fueled by curiosity and expressed through different mediums, forms the core of the art-science connection. It's a relationship built on the human impulse to explore, to make sense of the world around us, whether through a brushstroke or a scientific experiment. The act of observation itself, fundamental to both disciplines, fosters a deeper understanding, leading to new insights and innovative forms of expression.

Visualizing the unseen has always been a powerful driver of both scientific and artistic progress. Scientists employ sophisticated instruments to reveal hidden structures and processes, from the intricate machinery of cells to the vast expanse of the cosmos. Artists, in turn, take these revelations as inspiration, translating complex data and theories into tangible, visual forms. Imagine the swirling chaos of a nebula rendered in vibrant hues on canvas, or the delicate structure of a virus transformed into a glass sculpture. These artistic interpretations not only capture the beauty of scientific discoveries but also make them accessible to a wider audience, sparking curiosity and fostering a deeper appreciation for the wonders of the universe. They bridge the gap between abstract concepts and tangible experience.

The very act of inquiry, the process of asking questions and seeking answers, is central to both art and science. A scientist formulates hypotheses and designs experiments to test them. An artist explores different mediums, techniques, and perspectives to express their vision or to grapple with a particular theme. Both engage in a process of experimentation and refinement, driven by a desire to understand and communicate. Consider the Impressionist painters, who challenged conventional representations of light and color, or the pioneers of photography, who experimented with chemicals and light-sensitive materials to capture fleeting moments in time. This shared spirit of exploration, this constant push to challenge boundaries and push the limits of human understanding, is what truly unites art and science.

Bridging disciplines through the arts opens up exciting new avenues for scientific communication and public engagement. Complex scientific concepts, often obscured by jargon and mathematical formulas, can be made more accessible and engaging through visual and artistic mediums. Think of a dance performance choreographed to represent the intricate movements of subatomic particles or a musical composition inspired by the rhythmic patterns of brainwaves. These artistic interpretations can spark curiosity and understanding in audiences who might otherwise feel intimidated by scientific complexities. They offer a different entry point, an emotional connection, that can foster a deeper appreciation for the scientific world. The arts become a powerful tool for translating abstract ideas into tangible experiences.

The convergence of art and science offers a unique opportunity to challenge conventional ways of thinking and seeing the world. By combining the analytical rigor of science with the creative freedom of art, we can gain fresh perspectives and insights into complex issues. Imagine a collaborative project between a climate scientist and a sculptor, resulting in an installation that visually represents the impact of rising sea levels. Or consider a partnership between a neuroscientist and a musician, creating a sonic landscape that reflects the intricate workings of the brain. These collaborations not only enrich our understanding of scientific phenomena but also inspire new forms of artistic expression.

The future of art and science lies in embracing this interdisciplinary approach, fostering collaborations, and exploring new avenues for creative expression. As technology continues to evolve, it offers exciting possibilities for artists and scientists to push the boundaries of their respective fields. Imagine virtual reality experiences that allow us to explore the inner workings of a cell or interactive installations that respond to our brainwaves. These innovations have the potential to revolutionize how we engage with both art and science, creating immersive and transformative experiences that deepen our understanding of the world around us. It's a future where the lines between disciplines blur, and the fusion of art and science becomes a powerful catalyst for innovation and discovery. The intricate dance between art and science continues to evolve, driven by the shared human impulse to explore, understand, and create.

1.1 Bridging Disciplines

The act of bridging disciplines, especially those seemingly as disparate as art and science, requires more than simply placing them side by side. It demands a deep understanding of the inherent languages of both, recognizing their unique strengths and limitations. Think of it as learning to speak two distinct languages, not just memorizing vocabulary but grasping the nuances of grammar, the cultural context, and the unspoken subtleties. Only then can you begin to truly translate between them, conveying meaning accurately and effectively. This chapter will explore the approaches, methodologies, and philosophies that facilitate this cross-disciplinary dialogue, enabling artists to effectively interpret and express scientific concepts within their creative work. We'll examine how to build bridges, not just between art and science, but between the objective and the subjective, the tangible and the abstract.

One key approach lies in embracing the principles of visual thinking. Scientists use visual representations constantly – graphs, diagrams, models – to understand and communicate their findings. Artists, in turn, are trained to manipulate visual elements – color, composition, form – to convey emotions, narratives, and ideas. By recognizing this shared reliance on visual language, we can begin to see potential for translation. A scientist's graph, plotting the growth of bacteria, can become an artist's inspiration for a dynamic sculpture, capturing the explosive energy of life. A physicist's diagram of atomic structure might inform the intricate patterns in a textile design, revealing the hidden order within the seemingly chaotic. This translation process, however, requires careful consideration. Simply mimicking a scientific visual isn't enough; it's crucial to internalize the underlying concepts, allowing them to resonate with your own artistic vision.

Furthermore, the act of bridging disciplines often necessitates a shift in perspective. Scientists are trained to observe and analyze the world objectively, focusing on empirical data and quantifiable measurements. Artists, on the other hand, embrace subjectivity, exploring personal interpretations and emotional responses. Bridging these perspectives requires a willingness to embrace both the objective truth of scientific data and the subjective experience of artistic expression. Consider the intricate details revealed by a microscopic image of a diatom. A scientist might focus on classifying its structure, identifying its species, and analyzing its function within an ecosystem. An artist, however, might be drawn to its delicate symmetry, its translucent shell, or the ethereal play of light across its surface. By integrating both perspectives, the artist can create a work that not only captures the scientific accuracy of the diatom's structure but also evokes a sense of wonder and appreciation for its inherent beauty.

Collaboration plays a critical role in fostering these cross-disciplinary dialogues. Direct interaction between artists and scientists can provide invaluable insights and inspiration. Artists can gain access to cutting-edge research, advanced imaging technologies, and expert knowledge, while scientists can benefit from the artist's unique perspective, gaining new ways of visualizing and communicating their work. Imagine a sculptor collaborating with a biologist to create a three-dimensional representation of a virus. The biologist could provide detailed information about the virus's structure and function, while the sculptor could bring their expertise in form, material, and spatial relationships to create a tangible, visceral representation of this microscopic entity. This collaborative process not only enhances the artistic outcome but also deepens the scientist's understanding of their own research.

Developing a strong foundation in the relevant scientific principles is crucial for artists seeking to interpret scientific concepts effectively. This doesn't mean becoming a scientist, but rather acquiring a sufficient understanding of the underlying concepts to inform your artistic choices. If you're exploring quantum physics through your art, for instance, you need to grasp the fundamental principles of wave-particle duality, entanglement, and superposition. This understanding will provide a conceptual framework for your artistic explorations, allowing you to create work that is not only visually compelling but also conceptually grounded. Imagine an artist attempting to depict quantum entanglement without understanding the underlying principle. The resulting artwork might be aesthetically pleasing, but it would lack the intellectual depth and resonance that comes from a genuine engagement with the scientific concept.

Finally, it's essential to remember that the goal of bridging disciplines is not to illustrate scientific concepts literally but rather to interpret them creatively, using the language of art to explore their deeper implications and resonate with a broader audience. The artistic process should be a transformative one, where scientific concepts are not merely replicated but reimagined, becoming integrated into the artist's unique vision. A photograph of a nebula, for example, can be scientifically informative, but an artist's interpretation of that same nebula, perhaps through a painting or a sculpture, can evoke a sense of awe and wonder, prompting viewers to contemplate the vastness of the universe and humanity's place within it. This transformative power of art is what makes bridging disciplines so compelling, offering new ways of understanding and engaging with the world around us.

1.2   Visualizing the Unseen

The ability to visualize the unseen is a cornerstone of both scientific inquiry and artistic creation. Scientists utilize advanced instruments and technologies, from electron microscopes to radio telescopes, to peer into realms beyond our immediate perception. Artists, on the other hand, employ the tools of their craft—paint, clay, light, sound—to bring these unseen worlds to life, translating complex data and abstract theories into tangible, evocative forms. This chapter explores the diverse ways artists bridge the gap between the visible and invisible, transforming scientific discoveries into compelling visual narratives. We will delve into specific techniques artists use to represent the unseen, ranging from the meticulous accuracy of scientific illustration to the evocative abstraction of conceptual art.

Consider the intricate world revealed by microscopy. Once hidden, the delicate architecture of cells, the vibrant dance of microorganisms, are now accessible through powerful lenses. Artists take these magnified glimpses and reimagine them, employing diverse media to convey the wonder and complexity of this microscopic realm. Some artists meticulously reproduce the observed structures, creating detailed scientific illustrations that serve both aesthetic and educational purposes. Others draw inspiration from microscopic forms, transforming cellular structures into abstract sculptures or vibrant textile patterns. Imagine a large-scale installation composed of interwoven glass filaments, mimicking the intricate network of neurons, or a series of ceramic vessels whose surfaces echo the textured landscapes of bacterial colonies.

Beyond the microscopic, lies the even more elusive realm of quantum physics. Here, the very nature of reality becomes ambiguous, challenging our conventional understanding of space, time, and matter. Concepts like wave-particle duality, quantum entanglement, and superposition defy easy visualization. Yet, artists have found ingenious ways to represent these perplexing phenomena, utilizing metaphors and symbolism to convey their essence. Think of a painting where particles are depicted as both waves and points of light, simultaneously existing in multiple states, reflecting the principle of superposition. Or envision a sculpture where interconnected forms represent entangled particles, their fates inextricably linked despite vast distances, illustrating the concept of quantum entanglement.

The artistic interpretation of scientific concepts often involves a process of translation and transformation. Artists don't simply replicate scientific data; they interpret it, imbuing it with their own unique perspectives and artistic sensibilities. They might employ color palettes inspired by electron micrographs, creating vibrant, abstract paintings that evoke the energy and dynamism of cellular processes. They might use light and shadow to represent the interplay of forces within an atom, creating sculptures that capture the ephemeral nature of quantum phenomena. This process of artistic interpretation adds another layer of meaning to the scientific discovery, making it accessible and engaging for a wider audience.

Furthermore, the collaboration between artists and scientists can lead to new insights and discoveries. By visualizing scientific data in novel ways, artists can help scientists identify patterns and relationships that might have otherwise gone unnoticed. For instance, an artist's interpretation of complex protein folding patterns could inspire new avenues of research in biochemistry. Similarly, an artist's visualization of astronomical data could help astronomers better understand the structure and evolution of galaxies. These collaborations demonstrate the synergistic potential of art and science, highlighting how each discipline can enrich and inform the other.

As we explore the intersection of art and science, we begin to appreciate the power of visual language to communicate complex ideas. Art can illuminate the unseen, transforming abstract concepts into tangible forms, sparking curiosity, and fostering deeper understanding. Whether it is the intricate beauty of a diatom revealed through a microscope or the enigmatic dance of subatomic particles visualized through sculpture, artistic interpretations of science offer us a unique window into the hidden wonders of our universe. This interplay between art and science expands our perspectives, challenges our assumptions, and invites us to engage with the world in new and meaningful ways. It's a continuous dialogue, a constant exchange between observation, interpretation, and creation, pushing the boundaries of both artistic expression and scientific exploration. By embracing the power of visualization, we can unlock a deeper appreciation for the intricate beauty and complexity of the universe, from the smallest microbe to the vast expanse of the cosmos. This is the power of "Visualizing the Unseen."

1.3   The Art of Inquiry

Inquiry, in its purest form, is the engine driving both artistic and scientific progress. It is the insatiable curiosity, the burning question that propels us forward, pushing us to explore the unknown and make sense of the world around us. In the realm of art inspired by science, this inquiry takes on a unique dimension, bridging the gap between objective observation and subjective interpretation. For the artist engaging with scientific concepts, the process of inquiry involves not just understanding the scientific principles at play, but also finding ways to translate them into a visual language. This act of translation requires a deep dive into the subject matter, a willingness to grapple with complex ideas, and a persistent search for visual metaphors that can illuminate the underlying scientific truths.

Consider the artist exploring the microscopic world. Their inquiry might begin with observing micrographs of cellular structures, pondering the intricate architecture of life at its most fundamental level. This initial observation sparks questions: How can I capture the dynamic interplay of these cellular components? What colors and textures best convey the sense of constant motion and transformation within a living cell? How can I represent the unseen forces that govern these microscopic processes? These questions drive the artist's exploration, leading them to experiment with various artistic mediums and techniques. They might explore the fluidity of watercolor to depict the organic forms of cells, or the precision of etching to render the intricate details of their internal structures. The answers to these initial questions often give rise to new lines of inquiry, pushing the artistic process forward.

Similarly, for the artist engaging with quantum physics, the inquiry might begin with a fascination with concepts like entanglement or superposition. The artist might ask: How can I visualize something that exists in multiple states simultaneously? What artistic language can I use to express the interconnectedness of entangled particles? These abstract concepts, defying easy visualization, challenge the artist to think beyond conventional representation. They might explore the use of light and shadow to suggest the ephemeral nature of quantum phenomena, or create installations that invite the viewer to experience the disorienting nature of quantum reality. The very act of grappling with these complex ideas through an artistic lens can lead to new insights, both for the artist and the viewer.

The art of inquiry in this context involves not only asking the right questions but also cultivating a mindset of open exploration. It requires embracing ambiguity, accepting that some questions may not have definitive answers. It also necessitates a willingness to experiment, to try different approaches, and to learn from both successes and failures. Just as the scientist iteratively refines their experiments, the artist iteratively refines their creative process, constantly seeking new ways to express their understanding of the scientific concepts they are exploring. This iterative process, driven by inquiry, can lead to unexpected discoveries and breakthroughs, pushing the boundaries of both art and science.

In the fusion of art and science, the art of inquiry becomes a bridge between two seemingly disparate worlds. It is the common thread that connects the scientist's pursuit of objective truth with the artist's exploration of subjective experience. By embracing this spirit of inquiry, artists can not only illuminate complex scientific concepts but also inspire new ways of seeing and understanding the world around us. This process fosters a deeper appreciation for the interconnectedness of all things, from the smallest particles to the largest structures in the universe. Through this lens of artistic inquiry, science becomes not just a body of knowledge, but a source of wonder and inspiration. It is through this continuous cycle of questioning, exploring, and creating that we truly begin to unravel the mysteries of the universe, both within and without. This inquisitive approach transforms the artist into a visual storyteller, weaving narratives that connect the abstract world of scientific theory to the tangible realm of human experience.

2 The Microscopic World

The invisible world teems with life, a bustling metropolis of microorganisms, each with its own story. Consider the diatom, a single-celled algae encased in a glass shell of intricate, fractal patterns. These microscopic jewels, though individually invisible to the naked eye, collectively form vast oceanic blooms, influencing global climate and supporting marine ecosystems. Their delicate structures, viewed through the lens of a microscope, offer a wealth of inspiration for artistic exploration. Imagine translating the diatom's geometric precision into architectural designs, etched onto glass panels or woven into textile patterns. The iridescent shimmer of their silica shells could be captured in shimmering metallic paints or holographic installations, bringing the hidden beauty of the microscopic world to a macroscopic scale.

Beyond the diatom's crystalline architecture lies the dynamic world of bacteria. These single-celled organisms, often associated with disease, play vital roles in nutrient cycling, decomposition, and even human health. Their diverse shapes – from the spherical cocci to the rod-shaped bacilli and the spiral spirochetes – provide a rich visual vocabulary for artistic interpretation. Consider the potential of bacterial portraits: magnified images of individual bacterial cells, transformed into vibrant works of art through digital manipulation or hand-rendered illustrations. Their flagella, whip-like appendages used for movement, could be depicted as swirling brushstrokes or flowing lines of light, capturing their dynamic nature. The complex interactions within bacterial communities could be visualized as abstract compositions, reflecting the intricate networks and symbiotic relationships that govern their existence.

Exploring the microscopic realm requires specialized tools, foremost among them the microscope. This instrument, a portal to the invisible, allows us to witness the intricate details of cellular structures, revealing a hidden world of breathtaking complexity. The process of microscopy itself can be incorporated into the artistic process. Consider the technique of photomicrography, where images captured through a microscope are transformed into works of art. By manipulating lighting, focus, and magnification, artists can create ethereal landscapes from the microscopic world, highlighting the abstract beauty of cellular structures and biological processes. The use of fluorescent dyes can further enhance these images, adding vibrant colors and highlighting specific cellular components, creating stunning visual representations of life at the microscopic scale.

Beyond the purely visual, the microscopic world offers a wealth of conceptual inspiration. The concept of scale, for instance, can be explored through installations that juxtapose macroscopic representations of microscopic structures. Imagine a towering sculpture inspired by the structure of a virus, or a room-sized model of a cell, allowing viewers to walk through its internal architecture. These large-scale representations can offer a visceral understanding of the microscopic world, bridging the gap between the unseen and the tangible.

The very act of observing the microscopic world transforms our understanding of life. It reveals the interconnectedness of all living things, highlighting the intricate web of relationships that sustains life on Earth. This interconnectedness can be a powerful theme for artistic exploration, inspiring works that celebrate the diversity and interdependence of life at all scales. Consider an installation that uses interconnected networks of lights and sounds to represent the complex interactions within a microbial community. Or a series of paintings that depicts the lifecycle of a single cell, from its birth to its eventual division, highlighting the continuity of life across generations.

Furthermore, the microscopic world offers a unique perspective on the nature of life itself. It reveals the fundamental building blocks of all living organisms, highlighting the commonalities that unite all life on Earth. This fundamental unity can be a powerful source of inspiration, prompting artists to explore the essence of life through abstract representations of cellular structures and biological processes. Imagine a series of sculptures that depict the various stages of cell division, capturing the dynamic forces that shape life at its most fundamental level. Or a series of photographs that explores the intricate patterns found within cellular structures, revealing the hidden order and beauty that underlies all living things.

The exploration of the microscopic world also raises profound ethical questions. As our ability to manipulate life at the microscopic level increases, we must grapple with the implications of our actions. Art can play a crucial role in this ethical dialogue, prompting reflection on the responsibilities that come with our newfound power. Consider a series of paintings that depicts the potential consequences of genetic engineering, or a performance piece that explores the ethical dilemmas surrounding the use of CRISPR technology. By engaging with these complex issues, art can help us navigate the ethical challenges of the 21st century.

Finally, the microscopic world offers a powerful reminder of the vastness and complexity of the universe. It reveals a hidden world teeming with life, challenging our anthropocentric worldview and prompting us to consider our place within the larger cosmos. This expanded perspective can be a powerful source of inspiration, leading to artistic explorations that celebrate the interconnectedness of all things and the infinite possibilities of the universe.

2.1 Cellular Landscapes

The vibrant tapestry of a cell, invisible to the naked eye, presents a breathtaking landscape ripe for artistic exploration. Imagine the intricate network of the endoplasmic reticulum, a labyrinth of folded membranes, rendered in swirling brushstrokes of blues and greens, mimicking its fluid dynamism. Consider the Golgi apparatus, stacks of flattened sacs resembling an otherworldly cityscape, captured in a metallic sculpture, its surfaces reflecting the constant activity within. The possibilities for artistic interpretation are endless, each cellular component offering a unique form, texture, and function to inspire.

Think of the mitochondria, the powerhouses of the cell, pulsating with energy. A photographer could capture their essence using long-exposure techniques, transforming these organelles into radiant orbs of light, their movements tracing ethereal patterns across the image. A calligrapher, with deft strokes of ink, could depict the delicate dance of chromosomes during cell division, the elegant curves and loops mirroring the intricate choreography of life. A printmaker could layer textures and colors to represent the complexity of the cell membrane, a dynamic barrier selectively regulating the flow of substances in and out.

Moving beyond individual organelles, we can envision entire cellular landscapes. A painter might create a surrealist landscape depicting the bustling interior of a cell, with organelles morphing into fantastical structures, their interactions portrayed in a dreamlike narrative. A sculptor might craft a three-dimensional representation of a cellular environment, using different materials to represent the diverse components – glass for the transparency of the cytosol, polished stone for the rigidity of the nucleus, and woven fibers for the intricate cytoskeleton. This allows the viewer to engage with the cellular world in a tangible way, exploring its intricate architecture and the relationships between its parts.

Furthermore, the diverse array of cell types offers a rich palette for artistic expression. The branching dendrites of a neuron could be rendered in intricate wire sculptures, highlighting their complex interconnectedness. The contractile fibers of muscle cells could be represented through dynamic lines and shapes in a kinetic sculpture, conveying their rhythmic movements. The tightly packed cells of epithelial tissues could be depicted in mosaic form, each cell a distinct tile contributing to the larger pattern.

Artists can also explore the dynamic processes occurring within cells. Phagocytosis, the process by which a cell engulfs and digests foreign particles, could be represented through a series of photographs capturing the stages of engulfment, creating a visual narrative of this cellular drama. The process of protein synthesis, from DNA transcription to protein folding, could be depicted in a flowing, abstract painting, the colors and lines representing the complex molecular interactions involved.

Consider the use of airbrush techniques to create ethereal representations of cellular structures. The delicate gradation of color achievable with an airbrush can be used to depict the subtle variations in density within the cytoplasm or the intricate layering of membranes. This technique allows for a sense of depth and translucence, evoking the delicate nature of these microscopic structures. Moreover, artists can use their chosen medium to convey the functional significance of cellular components. For example, a sculptor creating a representation of the ribosome, the site of protein synthesis, might choose a material that reflects its role as a molecular machine – perhaps gears and levers interconnected to symbolize its dynamic function. A painter depicting the lysosome, the cell's waste disposal system, might use dark, brooding colors to evoke its role in breaking down cellular debris.

By delving into the intricate details of cellular structure and function, artists can create works that are not only visually stunning but also educational and insightful. These artistic interpretations can bridge the gap between scientific understanding and public perception, making the complex world of the cell accessible and engaging for a wider audience. Through their creative lens, artists can illuminate the hidden beauty and wonder of these microscopic landscapes, inspiring a deeper appreciation for the intricate machinery of life. The goal is to transform complex scientific concepts into accessible and engaging visual narratives.

2.2   Bacterial Portraits

Bacteria, often relegated to the realm of the invisible and undesirable, possess an unexpected aesthetic richness when viewed through an artistic lens. Think of them not as simple, uniform blobs, but as a diverse cast of characters, each with a unique morphology and a story to tell. Their vibrant colors, intricate textures, and surprising shapes offer a wealth of inspiration for artistic interpretation, revealing a hidden world of beauty and complexity. This chapter will delve into the techniques and approaches for creating compelling "bacterial portraits," transforming these microscopic organisms into captivating works of art.

One approach to capturing the essence of bacteria is through detailed scientific illustration. This method combines scientific accuracy with artistic flair, requiring careful observation and meticulous rendering. Begin by studying high-resolution micrographs, paying close attention to the specific features of each bacterial species. Note the arrangement of flagella, the texture of the cell wall, the clustering patterns of colonies. Translate these observations onto paper or a digital canvas, using fine lines, subtle shading, and precise detailing to create a scientifically accurate yet artistically expressive representation. Consider incorporating elements of design and composition to enhance the visual impact, perhaps by arranging different species in a dynamic interplay or highlighting specific structural details. The goal is to create a portrait that is both informative and visually engaging, inviting viewers to appreciate the intricate beauty of these microscopic life forms.

Moving beyond strict scientific representation, explore the potential of abstract interpretation. Consider the vibrant colors produced by certain bacterial pigments. Translate these hues into a swirling abstract painting, evoking the feeling of a bustling microbial community. Or perhaps focus on the rhythmic patterns formed by bacterial colonies, transforming them into a mesmerizing geometric design. Instead of directly depicting the bacteria themselves, use color, shape, and texture to evoke their presence and create an emotional response in the viewer. This approach allows for greater artistic freedom while still retaining a connection to the underlying scientific reality. Imagine a canvas awash in deep blues and greens, punctuated by vibrant streaks of yellow and orange, representing the complex interactions within a biofilm.

Photography offers another powerful medium for creating bacterial portraits. Macro photography, combined with specialized lighting techniques, can reveal the intricate details of bacterial colonies, transforming them into mesmerizing landscapes. Experiment with different lighting angles and colors to create dramatic effects and highlight specific textures. Consider using selective focus to draw attention to particular areas of interest, creating a sense of depth and dimension. Imagine a close-up shot of a bacterial colony, illuminated from the side to reveal its textured surface, resembling a miniature mountain range. The interplay of light and shadow can create a sense of drama and intrigue, drawing the viewer into the microscopic world.

Finally, consider incorporating mixed media techniques to further enhance your bacterial portraits. Combine scientific illustration with abstract painting, layering textures and colors to create a multi-dimensional representation. Incorporate found objects or natural materials that evoke the bacterial environment, such as sand, pebbles, or dried plant matter. Experiment with different printing methods, such as screen printing or etching, to create unique textures and patterns. Imagine a piece that combines a scientifically accurate illustration of a bacterium with abstract patterns created using bacterial pigments, layered onto a textured background made from agar, the substance used to culture bacteria. This multi-layered approach creates a rich and complex portrait, inviting viewers to explore the microscopic world from multiple perspectives. Through these artistic interpretations, we can transform the invisible world of bacteria into a source of wonder and inspiration, revealing the hidden beauty and complexity that lies beneath the surface. Remember, the goal is not simply to replicate what is seen under a microscope, but to interpret and transform it, creating a unique and evocative artistic statement.

3 Quantum Realms

The quantum world, a realm of probabilities and paradoxes, presents a unique challenge for artistic interpretation. Unlike the tangible forms of the microscopic world, quantum phenomena are often invisible and counterintuitive, defying easy visualization. Yet, this very elusiveness makes it a fertile ground for artistic exploration. Artists, much like scientists, grapple with uncertainty and ambiguity, seeking to make sense of the world through their respective mediums. In the quantum realm, where particles can be in multiple states at once and entanglement links the fates of distant objects, the artist's imagination can bridge the gap between abstract concepts and human experience.

Consider wave-particle duality, a cornerstone of quantum mechanics. A single quantum entity, like a photon or electron, can exhibit both wave-like and particle-like properties depending on how it is observed. This duality challenges our classical understanding of reality, where objects are neatly categorized as either waves or particles. Artists have responded to this conceptual puzzle by creating works that embrace ambiguity and fluidity. Some employ flowing, wave-like forms to represent the probabilistic nature of quantum entities, while others incorporate discrete, particle-like elements to suggest their quantized nature. The interplay between these contrasting visual elements mirrors the wave-particle duality itself, offering a tangible representation of an intangible concept.

Entanglement, another perplexing quantum phenomenon, presents a different set of artistic challenges. When two or more particles become entangled, their fates are intertwined, regardless of the distance separating them. Measuring the state of one instantly influences the state of the other, a connection that Einstein famously called "spooky action at a distance." Artists have explored this interconnectedness through various visual metaphors, such as interconnected lines, mirrored forms, or synchronized movements. Some create installations where interacting elements evoke the sense of instantaneous correlation between entangled particles, inviting viewers to contemplate the non-local nature of quantum reality. By transforming abstract mathematical relationships into tangible experiences, artists offer a visceral understanding of entanglement's implications.

Quantum tunneling, the ability of particles to pass through seemingly impenetrable barriers, further stretches our intuitive grasp of reality. This phenomenon, crucial for processes like nuclear fusion and radioactive decay, arises from the probabilistic nature of quantum particles. Instead of being confined by classical barriers, particles have a finite probability of "tunneling" through them. Artists have visualized this concept through imagery of particles disappearing and reappearing on the other side of barriers, or by depicting barriers as semi-transparent or permeable. Such representations, while simplified, convey the essence of quantum tunneling and its implications for the behavior of matter at the subatomic level.

Superposition, the ability of a quantum system to exist in multiple states simultaneously until measured, provides yet another avenue for artistic interpretation. Imagine a coin spinning in the air – it is neither heads nor tails until it lands. Similarly, a quantum particle in superposition exists in a combination of all possible states until observed. Artists have explored this concept through layered imagery, where multiple possibilities coexist within a single artwork. They might depict a particle as simultaneously occupying different locations, or represent a quantum system as a blend of various colors or shapes. These layered depictions capture the essence of superposition, highlighting the indeterminacy of quantum states before measurement.

Finally, the uncertainty principle, a fundamental limit on the precision with which certain pairs of physical properties can be simultaneously known, offers a powerful metaphor for the limits of human knowledge. Heisenberg's uncertainty principle states that the more precisely we know a particle's position, the less precisely we can know its momentum, and vice versa. This inherent uncertainty has been explored by artists through blurred imagery, fragmented forms, and works that embrace ambiguity. By acknowledging the limitations of our perception and understanding, these artistic interpretations of the uncertainty principle resonate with the inherent probabilistic nature of the quantum realm. They remind us that the act of observation itself can influence the reality we seek to understand. This dynamic interplay between observer and observed is not merely a scientific concept but a fundamental aspect of human experience, reflected in the very act of artistic creation.

3.1 Wave-Particle Duality in Art

The seemingly paradoxical nature of wave-particle duality presents a unique challenge and inspiration for artistic interpretation. How can one visually represent something that exists simultaneously as both a wave, spread out over space, and a particle, localized at a single point? This chapter explores the diverse ways artists have tackled this fundamental concept of quantum mechanics, transforming abstract scientific ideas into tangible, evocative works of art.

One approach artists employ involves the juxtaposition of contrasting visual elements. Imagine a painting where a single photon is depicted both as a concentrated point of light and as a ripple

expanding across a canvas. This simultaneous representation captures the duality inherent in the photon's nature, forcing the viewer
to confront the limitations of classical intuition. Another artistic
strategy involves using translucent layers or blurred forms to convey the wave-like aspect of quantum entities, while incorporating
sharp, defined points or objects to represent their particle-like behavior. This interplay of clarity and ambiguity mirrors the uncertainty inherent in quantum measurements.

Sculptural representations also offer intriguing possibilities. Consider a sculpture composed of both solid, discrete elements and
flowing, wave-like forms intertwined. Such a work could symbolize
the dual nature of matter, visually demonstrating how these seemingly contradictory aspects coexist. The use of materials like glass
or resin allows for exploration of transparency and light diffraction,
further enhancing the wave-like qualities of the artwork. Similarly,
kinetic sculptures that shift and change form can evoke the dynamic nature of quantum particles and their wave functions.

Photography, with its ability to capture both sharp detail and
ethereal blurs, can also be a powerful medium for exploring wave-
particle duality. Long-exposure photographs of light sources can
create wave-like patterns, while short exposures can freeze individual photons as distinct points. Combining these techniques in a
single image or series can create a compelling visual representation
of the duality. Furthermore, manipulating light through diffrac-

tion gratings or prisms in photographic compositions can produce interference patterns, directly visualizing wave-like behavior.

Digital art and animation provide even greater flexibility in representing this complex concept. Imagine an animation where a particle appears to dissolve into a wave as it travels through space, then re-materializes as a particle upon interaction with a detector. This dynamic transformation can visually capture the probabilistic nature of quantum phenomena, illustrating how the act of observation influences the system's behavior. Interactive digital installations can further engage viewers, allowing them to manipulate parameters and explore the interplay between wave and particle characteristics in real time.

Beyond these specific techniques, artists often draw inspiration from the philosophical implications of wave-particle duality. The concept challenges our understanding of reality, raising questions about the nature of observation and the limits of human perception. Some artworks explore these themes through abstract representations that evoke a sense of wonder and uncertainty. Others delve into the implications for our understanding of consciousness and the relationship between the observer and the observed.

The artistic exploration of wave-particle duality is not merely about visualizing a scientific concept; it's about grappling with profound questions about the nature of reality. Through their diverse interpretations, artists offer new perspectives on this fundamental aspect of quantum mechanics, making the invisible visible and inviting us to question our assumptions about the world around us. By translating complex scientific ideas into accessible and thought-provoking forms, artists bridge the gap between the abstract world of quantum physics and human experience. This fusion of art and science illuminates the profound mysteries of the universe and expands our understanding of the intricate interplay between the seen and the unseen. They transform complex equations and abstract theories into tangible, evocative experiences, sparking curiosity and fostering deeper engagement with the scientific world. Ultimately, these artistic explorations enrich both scientific understanding and artistic expression, demonstrating the powerful synergy that emerges when these two seemingly disparate disciplines converge.

3.2 Entanglement: Artistic Visions

The concept of entanglement, a cornerstone of quantum mechanics, presents a fascinating challenge for artistic interpretation. It describes a phenomenon where two or more particles become linked in such a way that they share the same fate, regardless of the distance separating them. Measuring the state of one instantaneously influences the state of the other, a connection that transcends our conventional understanding of space and time. Artists attempting to visualize this concept often grapple with the limitations of representing something fundamentally non-visual, a connection that exists beyond the realm of our everyday sensory experience. One approach involves exploring themes of interconnectedness and interdependence.

Imagine a sculpture composed of two seemingly disparate forms, perhaps crafted from contrasting materials like polished metal and rough-hewn wood. These forms, though distinct in their individual characteristics, are subtly linked by a network of fine threads or translucent wires, creating a visual representation of their shared destiny. The viewer is invited to contemplate the invisible forces that bind these objects together, mirroring the unseen connection between entangled particles. Another approach might involve the use of light and shadow. A projected image could depict two abstract shapes, their forms subtly mirroring each other. As the light source shifts, the shadows cast by these shapes intertwine and separate, symbolizing the fluctuating yet persistent connection between entangled particles. The ephemeral nature of light and shadow further emphasizes the elusive and counterintuitive nature of the quantum realm.

Artists can also explore the concept of entanglement through interactive installations. Imagine a pair of screens displaying constantly evolving patterns of color and light. The patterns on each screen, though seemingly random, are subtly correlated. When a viewer interacts with one screen, perhaps by touching it or moving in front of it, the patterns on both screens respond simultaneously, reflecting the instantaneous connection between entangled particles. This interactive element allows the viewer to directly engage with the concept of entanglement, experiencing its non-local nature firsthand. The use of sound can also play a powerful role in conveying the essence of entanglement. An audio installation might feature two distinct soundscapes, each evolving independently yet subtly influencing the other. The interplay of these soundscapes could create a sense of interconnectedness, mirroring the invisible link between entangled particles. Furthermore, the use of ambient sounds or subtle electronic manipulations can evoke the mysterious and otherworldly nature of the quantum realm.

Moving beyond traditional art forms, digital media offers exciting new possibilities for visualizing entanglement. Imagine a computer-generated animation depicting two particles moving through a virtual space. Their trajectories, though seemingly random, are subtly correlated. As one particle changes direction or speed, the other responds instantaneously, mirroring its movements across the virtual divide. The use of color, light, and motion in the digital realm can create visually stunning representations of entangled systems, allowing viewers to explore these complex concepts in a dynamic and engaging way. Another avenue for artistic exploration lies in the use of data visualization techniques. Scientists often use complex mathematical equations to describe the behavior of entangled particles. Artists can translate these equations into visual forms, creating abstract representations of the underlying mathematical relationships. This approach can reveal the inherent beauty and elegance of quantum mechanics, transforming complex scientific data into compelling works of art.

Photography also offers unique possibilities for exploring the concept of entanglement. Long-exposure photographs can capture the movement of light, creating ethereal and dreamlike images that evoke the mysterious nature of the quantum realm. By manipulating light and shadow, photographers can create visual metaphors for the interconnectedness of entangled particles, capturing the essence of this complex phenomenon in a single frame. Ultimately, the artistic interpretations of entanglement are as diverse as the artists themselves. Some artists may choose to focus on the abstract and theoretical aspects of the concept, creating works that explore the philosophical implications of quantum interconnectedness. Others may focus on the visual and aesthetic aspects, creating works that evoke the beauty and wonder of the quantum world. Regardless of their approach, artists play a crucial role in bridging the gap between science and the public, making complex scientific concepts accessible and engaging for a wider audience. Through their creative visions, they illuminate the hidden connections that bind our universe together, inviting us to contemplate the profound mysteries that lie at the heart of reality.

3.3   Quantum Tunneling: Expressions

Quantum tunneling, a phenomenon where particles defy classical physics by passing through energy barriers, offers a rich landscape for artistic interpretation. Imagine a rolling ball approaching a hill. Classically, if the ball doesn't have enough energy to reach the top, it simply rolls back. In the quantum world, however, there's a chance the ball could just "tunnel" straight through the hill, appearing on the other side. This seemingly magical act is not magic at all, but a direct consequence of the wave-like nature of particles at the quantum scale. The probability of this tunneling event decreases exponentially with the width and height of the barrier, but it's never truly zero. This concept, defying our everyday experience, opens doors to unique artistic expressions.

One can explore this concept visually by representing the wave function of a particle encountering a barrier. Instead of a solid, impenetrable wall, the barrier could be depicted as a translucent or semi-permeable field, allowing glimpses of the wave extending beyond it. The diminishing intensity of the wave within and beyond the barrier could be conveyed through color gradients, fading from vibrant hues to subtle shades, visually capturing the decreasing probability of finding the particle. Imagine a vibrant blue wave encountering a hazy purple barrier, the blue slowly fading within the purple but then reappearing, albeit fainter, on the other side.

This visual representation offers an intuitive grasp of the concept without resorting to complex mathematical equations.

Sculptural forms can also embody the essence of quantum tunneling. Consider a series of interconnected, undulating forms representing the particle's wave function. As the wave approaches a barrier, represented by a transparent or perforated material, parts of the sculpture could seamlessly pass through the openings, symbolizing the tunneling effect. The use of materials like glass or translucent resin would allow for the interplay of light and shadow, further enhancing the visual narrative of the tunneling process. This tangible representation allows viewers to interact with the concept spatially, experiencing the phenomenon from different perspectives.

Furthermore, the concept of tunneling can be incorporated into dynamic art forms like animation or digital installations. Visualize a particle as a pulsating point of light moving towards a barrier. Upon reaching the barrier, instead of being reflected, a portion of the light could subtly seep through, reforming on the other side. This visual transition could be accompanied by a corresponding auditory element, perhaps a subtle shift in pitch or tone, creating a multi-sensory experience that reinforces the concept of tunneling. The use of interactive elements, where viewers can manipulate the parameters of the simulation, such as the barrier's height or width, would further enhance engagement and understanding.

Another artistic avenue lies in exploring the implications of quantum tunneling in real-world phenomena. For instance, radioactive decay, a process where unstable atomic nuclei emit particles, is fundamentally governed by tunneling. Artists could create works that visually represent this process, showing particles escaping from the confines of the nucleus, defying the classical expectation of confinement. This could be portrayed through abstract imagery, using vibrant colors and dynamic forms to convey the energy released during the decay process. By linking the abstract concept of tunneling to a tangible physical process, the artwork can bridge the gap between the quantum realm and our everyday perception of reality.

The probabilistic nature of tunneling also lends itself to artistic exploration. Imagine an installation where multiple particles are represented by flickering lights. As these lights approach a barrier, some pass through while others are reflected, mirroring the probabilistic nature of tunneling. The frequency of tunneling events could be controlled by factors like barrier width, allowing viewers to witness the impact of these parameters on the outcome. Such an interactive installation could effectively communicate the statistical nature of quantum phenomena, a concept often challenging to grasp through traditional scientific representations. By intertwining artistic expression with scientific principles, we can unlock new ways of understanding and engaging with the profound mysteries of the quantum world. These artistic representations, though abstract, can serve as powerful tools for conveying the essence of quantum tunneling, making the invisible visible and the intangible tangible.

3.4 Superposition: Artful Depictions

Superposition, a cornerstone of quantum mechanics, presents a formidable challenge for artistic representation. It describes a quantum system existing in multiple states simultaneously until measured or observed, collapsing then into a single, definite state. Imagine a coin spinning in the air – it's neither heads nor tails until it lands. This "both/and" state, before it resolves into a definitive "either/or," captures the essence of superposition. How can artists capture this ephemeral, probabilistic reality in static forms?

One approach involves exploiting the inherent ambiguity within artistic mediums. Consider a sculpture composed of translucent materials, layering forms within forms. Depending on the angle of light and the viewer's perspective, different aspects of the sculpture are revealed or concealed. This shifting visibility mimics the uncertain, multifaceted nature of a quantum system in superposition. The artwork doesn't define a single, fixed reality, but instead offers a spectrum of potential interpretations, echoing the multiple states coexisting within the quantum realm. This approach utilizes light and shadow, translucency and opacity, to create a visual analogue for the superposition principle.

Another artistic strategy lies in embracing abstraction. Think of a painting composed of overlapping, semi-transparent brushstrokes of various colors. The colors blend and interact, creating new hues and patterns where they intersect. Each individual color represents a potential state of the quantum system, and their combination portrays the superposition of these states. The resulting image doesn't depict a single, definitive outcome but rather the dynamic interplay of possibilities, mirroring the probabilistic nature of quantum reality. This method leverages color theory and abstract expressionism to convey the complex essence of superposition visually.

Furthermore, artists can explore the concept of superposition through interactive installations. Imagine a room filled with suspended, shimmering particles that react to the viewer's presence. As the viewer moves through the space, these particles shift and rearrange, their patterns constantly evolving. This dynamic interaction symbolizes the act of measurement or observation in quantum mechanics, causing the superposition to collapse into a specific state. The artwork doesn't exist in a fixed form but is continuously reshaped by the observer's participation, reflecting the interconnectedness of observer and observed in the quantum world. Interactive art provides a powerful medium for engaging with the concept of superposition in a tangible, experiential way.

Time-based media like video art also offer unique possibilities for depicting superposition. A video could depict a single object morphing between multiple forms, transitioning seamlessly from one state to another. This constant flux and transformation visually captures the simultaneous existence of multiple states within a quantum system before measurement. The artwork unfolds over time, revealing the dynamic, evolving nature of superposition, rather than presenting a static representation. This approach utilizes the temporal dimension to convey the fluidity and indeterminacy inherent in the concept.

Finally, photography can be used to capture the essence of superposition through long exposures combined with intentional movement of the camera or subject. The resulting blurred image represents the multiple potential positions or states occupied by the subject during the exposure, mirroring the simultaneous existence of multiple states in a superposition. This technique leverages the photographic medium's ability to record time and motion to create a visual metaphor for the quantum phenomenon. The photograph doesn't capture a single, sharply defined moment but rather a spectrum of possibilities collapsed onto a two-dimensional plane.

By employing these diverse artistic strategies—exploiting ambiguity, embracing abstraction, utilizing interactive installations, incorporating time-based media, and leveraging the properties of photography—artists can effectively translate the complex and elusive concept of quantum superposition into tangible and thought-provoking artworks. These artistic interpretations not only offer visual metaphors for a scientific principle but also invite viewers to contemplate the fundamental nature of reality and the profound implications of quantum mechanics for our understanding of the universe. They bridge the gap between the abstract realm of quantum physics and the tangible world of human experience, fostering deeper engagement with one of the most intriguing concepts in modern science.

3.5   Uncertainty in Art & Science

Werner Heisenberg's uncertainty principle, a cornerstone of quantum mechanics, states that we cannot simultaneously know both the position and momentum of a particle with perfect accuracy. The more precisely we determine one, the less we know about the other. This inherent fuzziness at the subatomic level might seem far removed from the world of art, but a closer look reveals a surprising resonance. The act of artistic creation, particularly in abstract forms, often mirrors this principle of uncertainty. An artist working with a canvas, or even a digital space, may begin with a vague intention, a nebulous idea that takes shape gradually through the process of creation. Just as a physicist's attempt to pinpoint a particle's position can disrupt its momentum, so too can an artist's attempt to rigidly define their work prematurely stifle the creative flow.

Consider the act of observing a painting. Each viewer brings their unique experiences, biases, and interpretations to the piece. What one person sees as a vibrant expression of joy, another might perceive as a melancholic reflection on loss. Meaning, much like the quantum world, isn't fixed but rather emerges from the interaction between the observer and the observed. This inherent ambiguity allows for a multiplicity of readings, enriching the artistic experience and echoing the probabilistic nature of quantum phenomena. In photography, especially when exploring the microscopic, this principle manifests in the limitations of the medium itself. The diffraction of light, a wave-like property, places a fundamental limit on the resolution of an image. The sharper the focus on one detail, the more blurred its surroundings become, highlighting the inherent trade-off between precision and a broader perspective. This mirrors the uncertainty principle's constraints on simultaneously knowing both position and momentum. This limitation, however, can become a source of artistic expression, encouraging photographers to explore the interplay of light and shadow, clarity and ambiguity, to create images that transcend mere documentation and delve into the realm of artistic interpretation.

Similarly, in calligraphy, the deliberate imperfections, the subtle variations in ink flow and line weight, contribute to the overall aesthetic impact. These deviations from perfect uniformity, echoing the inherent uncertainty in the artist's hand and the fluidity of the medium, imbue the work with a sense of dynamism and life. The artist doesn't seek absolute control but rather embraces the element of chance, allowing the ink to guide the brush, much like a particle's unpredictable trajectory at the quantum level. Even airbrush art, known for its smooth gradients and seemingly precise control, grapples with a form of uncertainty. The interplay of air pressure, paint viscosity, and the artist's hand movements introduces an element of unpredictability, particularly when working with fine details or complex stencils. The artist must anticipate and respond to these subtle variations, constantly adjusting their technique to achieve the desired outcome, much like a physicist calibrating instruments to account for quantum fluctuations.

The uncertainty principle doesn't imply a lack of knowledge but rather a shift in how we understand knowledge itself. It encourages us to embrace ambiguity, to appreciate the multiplicity of perspectives, and to recognize that meaning is not always fixed but can emerge from the interplay of different elements. This shift in perspective resonates deeply with the artistic process, where ambiguity and uncertainty often serve as catalysts for creativity. The artist, like the quantum physicist, navigates a landscape of possibilities, embracing the unknown and transforming it into something meaningful and expressive.

Furthermore, the uncertainty principle highlights the interconnectedness of observer and observed. The act of observation influences the system being observed, blurring the lines between subject and object. This interrelationship finds parallels in art, where the viewer's engagement with a piece becomes an integral part of its meaning. The artwork exists not in isolation but in a dynamic relationship with its audience, inviting interpretation and co-creation of meaning. The viewer doesn't passively consume the art but actively participates in its unfolding.

Uncertainty, then, is not a barrier to understanding but a gateway to deeper exploration, both in science and art. It reminds us that the universe, like a work of art, is not a static entity but a dynamic process, constantly evolving and revealing new layers of meaning. It is in embracing this uncertainty, in accepting the limitations of our knowledge, that we can truly appreciate the beauty and complexity of the world around us, both at the macroscopic and microscopic levels.

4   Art Meets Microscopy

The marriage of art and microscopy offers a unique opportunity to explore the invisible landscapes hidden within our world. Microscopy, with its ability to magnify the minute, reveals intricate details and structures otherwise imperceptible to the naked eye. This visual data, rich in texture, form, and pattern, provides a potent source of inspiration for artistic expression. Consider the ethereal beauty of diatoms, single-celled algae encased in intricate glass shells. Their delicate, symmetrical forms, revealed through the lens of a microscope, have captivated scientists and artists alike, inspiring intricate drawings, sculptures, and even architectural designs. This chapter delves into the ways artists transform micrographs, those captivating images produced by microscopes, into compelling works of art.

Beyond simply replicating the visual information provided by a micrograph, artists interpret and transform these scientific images, imbuing them with personal meaning and artistic vision. They might isolate and amplify specific details, playing with color, contrast, and composition to create abstract representations of microscopic structures. Imagine a painting inspired by the branching network of neurons, where the artist uses vibrant hues and flowing lines to evoke the dynamic energy of the nervous system. Or envision a sculpture based on the crystalline structure of a virus, transforming the cold, hard data of science into a tangible, three-dimensional form. The possibilities are as vast and diverse as the microscopic world itself.

One fascinating avenue of artistic exploration involves "sculpting the invisible." Using techniques like 3D printing and micromachining, artists can create physical representations of microscopic structures, allowing viewers to interact with these normally unseen worlds in a tangible way. Imagine holding a model of a pollen grain, enlarged thousands of times, and feeling its textured surface. This tactile experience adds another dimension to our understanding of the microscopic realm, bridging the gap between scientific observation and sensory perception. Artists might also incorporate materials that reflect the properties of the microscopic structures they represent, such as using translucent resins to mimic the delicate membranes of cells. This careful selection of materials adds depth and meaning to the artwork, enriching the viewer's experience.

Another innovative approach involves the use of microbes themselves as artistic mediums. "Painting with microbes" is a technique where artists cultivate bacteria, fungi, or other microorganisms on a growth medium, using their natural pigments and growth patterns to create living artworks. These "microbial masterpieces" evolve over time, as the microorganisms multiply and interact, introducing an element of unpredictability and change into the artistic process. The resulting patterns and colors can be breathtaking, offering a unique blend of science and art. Think of a petri dish transformed into a canvas, where vibrant colonies of bacteria create intricate designs, a testament to the hidden beauty and complexity of microbial life.

The intersection of art and microscopy transcends mere aesthetics; it fosters deeper understanding and appreciation for the scientific world. By translating complex scientific data into visually compelling artworks, artists make the invisible visible, engaging audiences in a way that traditional scientific communication often cannot. These artistic interpretations spark curiosity, prompting viewers to ask questions about the underlying science and to explore the microscopic world with a newfound sense of wonder. They challenge us to reconsider our relationship with the unseen, reminding us that even the smallest of organisms play a vital role in the intricate web of life. Furthermore, this fusion of disciplines encourages dialogue and collaboration between artists and scientists, leading to new insights and perspectives. This interdisciplinary approach not only enriches the artistic and scientific fields but also fosters a greater appreciation for the interconnectedness of knowledge and the power of creative exploration. It illuminates the beauty and complexity of the natural world, inviting us to marvel at the hidden wonders that surround us.

4.1   Micrographs as Inspiration

The beauty of a micrograph lies not just in its scientific value, but also in its inherent artistry. The intricate patterns, textures, and colors revealed under the microscope can be as captivating as any painting or photograph. These images offer a unique glimpse into a hidden world, revealing the breathtaking complexity of life at the smallest scales. For an artist, micrographs provide a rich source of inspiration, a starting point for creative explorations that bridge the gap between science and art. Imagine the swirling patterns of diatoms, their silica shells etched with intricate, geometric designs. These microscopic algae, invisible to the naked eye, become exquisite works of art when viewed through the lens of a microscope. Their delicate structures can inspire the creation of intricate jewelry, the design of patterned textiles, or even the architecture of buildings.

Consider the vibrant colors of stained tissue samples. Pathologists use these stains to differentiate cellular structures, but the resulting images are often strikingly beautiful, resembling abstract expressionist paintings. The contrast between healthy and diseased tissues can create a powerful visual narrative, prompting reflection on the fragility of life and the intricate workings of the body. An artist might be drawn to the interplay of colors and shapes, translating the microscopic world into a large-scale artwork that evokes both wonder and contemplation.

Beyond the purely aesthetic appeal, micrographs offer a window into the fundamental principles of life. They reveal the building blocks of organisms, the complex interactions between cells, and the dynamic processes that drive biological systems. Observing the branching networks of neurons, the rhythmic pulsing of cilia, or the intricate dance of chromosomes during cell division can be a profound experience. These observations can inspire artistic interpretations that explore themes of growth, interconnectedness, and the cyclical nature of life. A sculptor might be moved to create a three-dimensional representation of a neuronal network, capturing its intricate branching structure and highlighting its role in information processing.

To effectively use micrographs as inspiration, it's crucial to understand the underlying science. Knowing the function of a particular structure or the process being depicted can add depth and meaning to an artwork. For instance, understanding the role of mitochondria in energy production might inspire an artist to create a work that visually represents the flow of energy within a cell. This scientific grounding can inform the choice of materials, colors, and composition, resulting in a more meaningful and impactful artwork. One could envision a kinetic sculpture powered by solar energy, mimicking the energy conversion process within a mitochondrion. Exploring different microscopy techniques can also open up new artistic possibilities. Electron microscopy, for example, reveals details far beyond the reach of traditional light microscopy, offering images of cellular ultrastructure with remarkable clarity. Confocal microscopy allows for the creation of three-dimensional reconstructions of specimens, offering a unique perspective on their spatial organization. By understanding the capabilities and limitations of different microscopy techniques, artists can make informed choices about which images best suit their artistic vision. A digital artist could manipulate confocal microscopy data to create a virtual reality experience that allows viewers to explore the inner workings of a cell.

Collaboration between artists and scientists can further enrich the creative process. Scientists can provide access to specialized equipment and expertise, while artists can offer fresh perspectives and innovative ways of interpreting scientific data. These collaborations can lead to the development of new artistic techniques and the creation of works that effectively communicate complex scientific concepts to a wider audience. Imagine an artist working with a biologist to create a series of paintings based on time-lapse microscopy of developing embryos, documenting the intricate process of cell differentiation and tissue formation.

Finally, remember that the goal is not simply to replicate the micrograph. Instead, use it as a springboard for your own creativity. Experiment with different artistic media, explore different styles, and don't be afraid to abstract and interpret the imagery in your own unique way. The micrograph is a starting point, a source of inspiration that can lead to a wide range of artistic expressions. Let the intricate beauty of the microscopic world ignite your imagination and guide you on a journey of artistic discovery. By bridging the gap between science and art, we can gain a deeper appreciation for the wonders of the universe, from the smallest scales to the largest. The marriage of these two disciplines holds immense potential for fostering creativity, expanding our understanding of the world, and inspiring new ways of seeing.

4.2   Sculpting the Invisible

The tangible world often captivates us with its readily apparent forms, but an entire universe exists beyond our immediate perception. Within the microscopic realm, intricate landscapes and hidden dramas unfold, invisible to the naked eye. This chapter explores how artists can "sculpt the invisible," transforming microscopic imagery and biological concepts into tangible artistic expressions. We will journey beyond mere representation, venturing into a realm where art breathes life into scientific data, forging a deeper connection between the viewer and the unseen world.

Consider the intricate architecture of diatoms, single-celled algae with glass-like shells displaying remarkable geometric patterns. These microscopic marvels offer a rich tapestry of forms, inspiring artists to recreate their delicate structures in larger-than-life sculptures. Imagine a glassblower meticulously crafting a diatom's ornate shell, magnifying its intricate details to reveal a hidden world of breathtaking complexity. Or envision a sculptor using metal wire to capture the delicate tracery of a radiolarian's skeleton, transforming a microscopic organism into a mesmerizing, three-dimensional form. These artistic interpretations not only showcase the beauty of these tiny organisms but also invite viewers to contemplate the intricate designs found throughout nature, prompting reflection on the underlying scientific principles that govern their formation.

Beyond individual organisms, artists are also exploring the vibrant ecosystems thriving within a single drop of water. Imagine a large-scale installation representing a microbial community, where different bacterial colonies are depicted using various materials and textures. Biofilms, complex communities of microorganisms, could be rendered as intricately layered sculptures, revealing the dynamic interactions between different species. By visualizing these normally invisible processes, artists can convey the vital roles microorganisms play in our environment and their impact on human health.

This artistic lens can also explore the impact of environmental factors on microbial communities, showcasing how pollution or climate change can disrupt these delicate ecosystems.

The process of sculpting the invisible often involves a close collaboration between artists and scientists. Artists can leverage microscopy techniques, such as scanning electron microscopy (SEM) and confocal microscopy, to capture high-resolution images of microscopic structures. These images then serve as blueprints for their artistic creations, allowing them to faithfully recreate the intricate details of the microscopic world. Furthermore, consultations with biologists and other scientists provide valuable insights into the biological processes and underlying scientific principles being visualized, enriching the artistic interpretation with a layer of scientific accuracy.

Consider an artist collaborating with a microbiologist to create a series of sculptures depicting the life cycle of a virus. The artist might use translucent materials to represent the virus's outer shell and incorporate internal structures based on scientific data obtained through electron microscopy. Different stages of the viral life cycle, from attachment to replication and release, could be captured in a series of interconnected sculptures, creating a narrative that unfolds in three dimensions. This collaborative approach not only results in visually compelling artwork but also serves as a powerful tool for science communication, making complex biological concepts accessible to a wider audience.

Moreover, sculpting the invisible extends beyond mere replication of microscopic forms. Artists can use their creative license to interpret and abstract scientific data, creating works that evoke emotions and spark curiosity. Imagine an artist using light and shadow to represent the dynamic fluctuations of protein molecules within a cell, creating an ethereal and evocative installation that captures the essence of cellular processes. Or consider an artist using vibrant colors and fluid forms to depict the flow of energy through an ecosystem, transforming scientific data into an abstract yet captivating visual narrative.

By transforming the unseen into tangible forms, artists bridge the gap between the scientific and artistic realms, inviting viewers to engage with scientific concepts in new and meaningful ways. These artistic interpretations not only enhance our understanding of the microscopic world but also inspire a sense of wonder and appreciation for the hidden beauty and complexity that surrounds us. They challenge us to look beyond the surface and explore the vast, invisible universe that shapes our world in profound ways. The act of sculpting the invisible thus becomes a powerful act of discovery, both for the artist and the viewer, forging a deeper connection between science, art, and the human experience.

4.3 Painting with Microbes

Agar art is a captivating intersection of art and microbiology, transforming petri dishes into canvases and bacteria into living pigments.
This technique allows artists to cultivate vibrant, intricate designs
using the natural growth patterns of microorganisms. Imagine
painting not with brushes and oils, but with colonies of Serratia
marcescens, their crimson hues blooming across the agar surface.
Or perhaps employing the golden yellow of Micrococcus luteus to
create shimmering landscapes within the confines of a dish. The
possibilities are as diverse as the microbial world itself.

Preparing for microbial art requires a sterile environment and a
few essential supplies. Nutrient agar, a jelly-like substance that
provides sustenance for bacteria, serves as the canvas. Sterilized
petri dishes contain the agar and provide a controlled environment
for growth. You'll also need an inoculation loop, a small wire loop
used to transfer bacteria onto the agar. Finally, and most impor-
tantly, you'll need a selection of pigmented bacteria. These can be
obtained from biological supply companies or, with proper training
and caution, cultured from environmental samples.

Selecting the right bacteria is crucial for achieving desired artistic
effects. Different species produce a spectrum of colors, from the
deep violets of Chromobacterium violaceum to the creamy whites
of Bacillus subtilis. Researching the pigmentation and growth char-

acteristics of various bacteria is essential for planning your composition. Consider how colors might blend or contrast, and how the varying growth rates of different species could contribute to the final artwork's texture and form. A slower-growing bacterium might provide fine details, while a faster-growing one could create bold, sweeping strokes of color.

The actual painting process involves carefully transferring the chosen bacteria onto the agar surface. Using a sterile inoculation loop, a small amount of bacteria is picked up from a pure culture and gently streaked onto the agar in the desired pattern. Precision and a steady hand are key, especially for intricate designs. Think of it as drawing with invisible ink, the final image only revealed as the bacteria multiply and their colors emerge. Varying the density of the bacterial streaks can create gradients of color and texture, adding depth to the artwork. A lighter touch might yield translucent washes of color, while a denser application could produce opaque, vibrant hues.

Incubation is the next crucial step. The inoculated petri dishes are placed in a temperature-controlled incubator, typically set at a temperature optimal for bacterial growth. Over several days, the bacteria multiply, their colonies expanding and their pigments becoming visible. This process transforms the initially blank agar into a living artwork, the colors and patterns evolving over time. Regular observation allows you to monitor the growth and capture the dynamic changes in your microbial masterpiece. You might choose to stop the growth at a particular stage by refrigerating the dish, preserving the artwork at its peak.

Safety precautions are paramount when working with microorganisms. Even though the bacteria used in agar art are generally considered safe, maintaining a sterile environment and following proper handling procedures are essential. Always work in a clean space, sterilize your equipment thoroughly, and avoid contaminating your cultures. Dispose of used petri dishes responsibly, following established biohazard guidelines. Treat all microbial cultures with respect, acknowledging their living nature and potential risks.

Beyond the aesthetic appeal, agar art offers a unique opportunity to explore the intersection of art and science. It's a powerful tool for visualizing the invisible world of microbes, transforming them from abstract concepts into tangible, vibrant forms. This process can spark curiosity about the microbial world and inspire further exploration of its complexities. Experimenting with different bacteria, agar types, and incubation conditions opens up endless possibilities for artistic expression, pushing the boundaries of this fascinating medium. Consider incorporating elements of traditional art techniques, such as pointillism or abstract expressionism, into your microbial designs.

Finally, documenting your microbial artwork is essential. Photography is a natural choice, allowing you to capture the vibrant colors and intricate patterns. Experiment with different lighting and backgrounds to highlight the unique textures and three-dimensional qualities of your microbial creations. Time-lapse photography can also be a compelling way to document the growth process, revealing the dynamic evolution of the artwork over time. Sharing your microbial art with others can spark conversations about the intersection of art and science, fostering a deeper appreciation for the beauty and complexity of the microbial world. Consider participating in agar art competitions or exhibitions, or simply sharing your creations online to inspire others to explore this captivating medium.

5 Art Inspired by Quantum Physics

Quantum physics, with its paradoxical principles and mind-bending concepts, presents a unique challenge and inspiration for artists. This realm, where particles can be in multiple states at once and entanglement links the fates of distant objects, defies our everyday experiences and demands new forms of visual expression. Exploring this intersection between quantum mechanics and artistic creation, we discover a fertile ground for innovation and a powerful lens through which to examine the nature of reality. Artists venturing into this territory are not simply illustrating scientific concepts; they are grappling with profound philosophical questions about the universe and our place within it.

The inherent abstraction of quantum phenomena lends itself naturally to artistic interpretation. Concepts like superposition, where a quantum particle exists in all possible states until measured, can be visually represented through layered imagery, translucent forms, or works that evolve over time, reflecting the shifting probabilities of the quantum world. Imagine a sculpture composed of shimmering, interconnected elements, constantly shifting and rearranging, embodying the dynamic nature of quantum states. Or perhaps a series of photographs, each capturing a slightly different configuration of light and shadow, hinting at the multiple potentialities contained within a single quantum system.

Entanglement, another cornerstone of quantum mechanics, offers equally compelling artistic possibilities. This phenomenon, where two or more particles become linked and share the same fate regardless of distance, can be explored through interconnected artworks, mirrored patterns, or installations that respond to each other in real-time. Consider two canvases, separated by a significant distance, where brushstrokes on one instantaneously manifest on the other, a visual metaphor for the interconnectedness of entangled particles. Or a sound installation where tones and rhythms generated by one instrument subtly influence the sounds produced by another, creating an auditory representation of quantum entanglement.

Beyond these specific concepts, the broader philosophical implications of quantum physics provide rich material for artistic exploration. The uncertainty principle, which states that certain pairs of physical properties cannot be simultaneously known with perfect accuracy, can inspire works that embrace ambiguity and chance.

Imagine a painting created using random drips and splashes of color, where the final form emerges from the interplay of chance and intention, mirroring the inherent uncertainty at the heart of quantum mechanics. Or a performance piece where the actions of the performers are determined by the outcome of quantum measurements, introducing an element of unpredictability and challenging our notions of cause and effect.

Furthermore, the very act of observing a quantum system influences its behavior, a concept known as wave function collapse. This observer effect can be explored through interactive art installations where the viewer's presence or actions directly shape the artwork, highlighting the interconnectedness between observer and observed. Imagine a projection that changes based on the viewer's gaze, or a sculpture that responds to the viewer's touch, reflecting the active role of the observer in shaping quantum reality.

Creating art inspired by quantum physics often requires a deep engagement with the scientific principles involved. Artists may collaborate with physicists, study scientific literature, or even conduct their own experiments to gain a deeper understanding of the concepts they are exploring. This cross-disciplinary approach enriches both the artistic and scientific realms, fostering new ways of thinking and seeing.

The artistic exploration of quantum physics is not merely an intellectual exercise; it is a powerful means of communicating complex scientific ideas to a wider audience. By translating abstract concepts into tangible forms, artists can make the invisible world of quantum mechanics accessible and engaging. These artworks serve as a bridge between the esoteric world of scientific research and the lived experience of the viewer, sparking curiosity and inspiring further exploration.

This fusion of art and quantum physics represents a new frontier in both disciplines. As artists continue to grapple with the profound implications of quantum mechanics, we can expect to see even more innovative and thought-provoking works that challenge our perceptions of reality and expand our understanding of the universe. This ongoing dialogue between art and science promises to enrich both fields, fostering new insights and inspiring new generations of artists and scientists. Through their unique perspectives and creative visions, artists are illuminating the hidden depths of the quantum realm, offering us a glimpse into the fundamental nature of reality and our place within this strange and wonderful universe. This exploration continues to evolve, blurring the lines between science, philosophy, and art, and prompting us to question what we think we know about the world around us. The quantum world, once confined to the realm of scientific inquiry, is now finding expression in the vibrant and evolving landscape of contemporary art.

5.1 Quantum Art: A New Frontier

Quantum art isn't merely about illustrating scientific concepts; it's about grappling with the philosophical and aesthetic implications of a reality we can barely grasp. It's about translating the invisible language of quantum mechanics into a sensory experience, using the tools of art to bridge the gap between abstract theory and human perception. This new frontier pushes the boundaries of both art and science, demanding a unique blend of creative intuition and scientific literacy. Think of it as a dialogue, not a lecture. Quantum physics poses questions about the very nature of reality, questions that have resonated with artists for centuries.

Consider the concept of wave-particle duality, the idea that quantum entities can exhibit both wave-like and particle-like behavior. How can something be two seemingly contradictory things at once? This paradox has captivated artists, inspiring works that explore the interplay of form and formlessness, presence and absence. Some artists use light and shadow to represent this duality, creating installations where photons behave as both waves and particles, blurring the lines between what we perceive and what actually is. Others employ sculpture, crafting forms that shift and change depending on the viewer's perspective, mirroring the elusive nature of quantum objects.

Entanglement, another mind-bending principle of quantum mechanics, presents a unique challenge and opportunity for artistic interpretation. This phenomenon describes a connection between two or more quantum particles, where their fates are intertwined regardless of the distance separating them. Artists have explored this concept through interconnected installations, where the manipulation of one element instantaneously affects another, creating a sense of invisible linkage. Performance art can also capture the essence of entanglement, with dancers mirroring each other's movements, symbolizing the interconnectedness of quantum entities. Imagine a canvas where brushstrokes on one side mysteriously manifest on the other, a visual representation of this "spooky action at a distance," as Einstein famously called it.

Quantum tunneling, the seemingly impossible ability of a particle to pass through a barrier it doesn't have the energy to overcome, provides yet another avenue for artistic exploration. Sculptures that appear to defy gravity, paintings where colors bleed through seemingly impenetrable boundaries, and digital art where forms materialize and dematerialize across virtual barriers can all evoke the uncanny nature of this quantum phenomenon. Think of it as visually representing the probabilistic nature of reality, where the seemingly impossible can and does occur.

Superposition, the principle that a quantum system can exist in multiple states simultaneously until measured, offers fertile ground for artistic interpretation. Imagine a painting that shifts and changes depending on the viewer's gaze, or a sculpture that embodies multiple forms at once, reflecting the superposition of quantum states. Artists can use layered images, shifting perspectives, and interactive elements to create works that embody the multifaceted nature of quantum reality. This invites viewers to question their own role in defining what they observe, echoing the observer effect in quantum mechanics.

The uncertainty principle, which states that there's a fundamental limit to the precision with which certain pairs of physical properties of a particle can be known, challenges the very notion of objective observation. Artists can embrace this uncertainty through abstract works that defy precise interpretation, inviting viewers to engage with the ambiguity and embrace the inherent limitations of our knowledge. Think of hazy, dreamlike landscapes, or sculptures that resist clear definition, mirroring the inherent fuzziness of the quantum world.

Creating quantum art requires more than just an understanding of the scientific concepts. It demands a willingness to embrace ambiguity, to question our assumptions about reality, and to translate complex ideas into a language that resonates with human experience. It's about creating a sense of wonder, sparking curiosity, and prompting viewers to contemplate the profound mysteries that lie at the heart of existence. It's a journey of discovery, where the intersection of art and science illuminates not only the quantum world, but also the nature of creativity itself. This is the true power of quantum art: to transform the abstract into the tangible, the invisible into the visible, and the incomprehensible into a source of wonder and inspiration.

5.2   Abstraction & Quantum Reality

The quantum world, with its inherent uncertainties and paradoxical behaviors, presents a unique challenge for artistic representation. Unlike the tangible forms of the microscopic world, quantum phenomena exist in a realm beyond direct observation, governed by probabilities and wave functions rather than concrete structures. This very intangibility, however, opens up exciting possibilities for artistic exploration, inviting artists to delve into abstraction as a primary tool for conveying the essence of quantum reality. Abstraction, in this context, moves beyond mere aesthetic choice; it becomes a necessary language for expressing the fundamental unknowability and strangeness that lie at the heart of quantum mechanics.

Consider the concept of superposition, where a quantum particle can exist in multiple states simultaneously until measured. How can one visually depict such a counterintuitive notion? Traditional representational art, with its focus on depicting observable reality, falls short. Instead, artists might employ layered imagery, translucent forms, or ambiguous shapes that evoke a sense of multiplicity and potentiality. Color can also play a powerful role, with shifting hues and overlapping tones suggesting the simultaneous presence of different states. The canvas becomes a space not for replicating reality, but for embodying the very idea of superposition itself.

Similarly, the principle of entanglement, where two or more particles become linked and share a fate regardless of distance, defies easy visualization. Artists might explore this concept through interconnected forms, mirrored patterns, or lines that visually tie together disparate elements within a composition. The goal is not to illustrate the mechanism of entanglement, which remains a mystery even to physicists, but to capture the sense of interconnectedness and non-locality that it implies. This artistic interpretation can be achieved through various techniques, including digital manipulation and mixed media installations that incorporate elements of light and sound.

Quantum tunneling, the phenomenon where a particle can pass through a barrier seemingly impenetrable according to classical physics, presents another intriguing challenge. Artists might represent this by depicting objects phasing through seemingly solid boundaries, using blurred edges or ghostly outlines to suggest the particle's ethereal passage. Alternatively, they could employ visual metaphors, such as depicting a river flowing through a mountain, to convey the idea of traversing an obstacle that appears insurmountable. The choice of medium also becomes crucial here; sculptures incorporating transparent materials or installations using projected light could effectively communicate the immateriality and surprising nature of this quantum behavior.

The uncertainty principle, which dictates that certain pairs of physical properties, like position and momentum, cannot be simultaneously known with perfect accuracy, lends itself to artistic expressions of ambiguity and indeterminacy. Blurred lines, hazy forms, and incomplete figures can convey the inherent limitations on our knowledge of a quantum system. The use of negative space, areas intentionally left blank or undefined, can also powerfully represent the unknown and unknowable aspects of quantum reality. By embracing these techniques, artists can invite viewers to engage with the uncertainty inherent in the quantum world, fostering a sense of wonder and intellectual curiosity.

Furthermore, the wave-particle duality, where quantum entities exhibit both wave-like and particle-like properties depending on the experimental setup, offers rich ground for artistic exploration. Artists might juxtapose smooth, flowing lines representative of waves with sharp, defined shapes suggestive of particles, creating a visual tension that reflects the dual nature of quantum objects. They could also use dynamic patterns and rhythmic variations in color and texture to evoke the sense of wave propagation and particle interaction.

Ultimately, abstraction in quantum art becomes a means of translating complex scientific concepts into a language accessible to a wider audience. It allows artists to create visual metaphors that capture the essence of quantum phenomena without resorting to oversimplification or inaccurate depictions. By embracing the ambiguity and uncertainty inherent in the quantum realm, artists can stimulate dialogue and encourage viewers to grapple with these profound ideas in new and imaginative ways. The result is a form of art that not only illuminates scientific principles but also expands our understanding of reality itself. Through the creative use of color, form, and space, artists can convey the mysterious and paradoxical nature of quantum phenomena, prompting viewers to contemplate the fundamental nature of the universe and our place within it.

6 The Future of Art-Science Fusion

The convergence of art and science is poised for a period of dynamic growth, fueled by technological advancements, expanding interdisciplinary dialogue, and a growing recognition of their combined power. This future isn't just about creating aesthetically pleasing representations of scientific concepts; it's about fostering a deeper understanding and appreciation of the universe around us, from the subatomic to the cosmic. It's about sparking curiosity, igniting new research avenues, and bridging the gap between complex scientific discoveries and public perception. This fusion promises to revolutionize not just how we perceive science, but how we engage with it.

One crucial aspect of this evolution is the burgeoning field of interdisciplinary collaboration. Scientists and artists are increasingly recognizing the value of working together, sharing their unique perspectives and skillsets to create truly innovative works. Scientists can provide artists with access to cutting-edge research, data visualization tools, and specialized equipment like advanced microscopes, while artists can offer fresh perspectives, visual storytelling expertise, and the ability to translate complex data into accessible and engaging forms. These collaborations are breaking down traditional disciplinary silos, leading to a richer, more nuanced exploration of the intersection between art and science. Think of an artist working with a biophysicist to create a sculptural installation based on protein folding or a choreographer collaborating with a quantum physicist to develop a dance performance interpreting the concept of entanglement. These partnerships aren't just about creating art; they're about forging new pathways for discovery and understanding.

The development of new media and technologies plays a pivotal role in this ongoing fusion. Virtual reality, augmented reality, 3D printing, bio-art, and data sonification offer artists unprecedented opportunities to explore and interpret scientific concepts in dynamic and interactive ways. Imagine exploring a virtual reality simulation of a cell's interior, experiencing the intricate dance of molecules firsthand. Or consider a 3D-printed sculpture generated from the data of a particle collider experiment, translating the invisible realm of subatomic particles into a tangible, aesthetically compelling form. These technologies empower artists to create immersive experiences that deepen public engagement with scientific principles, making abstract concepts tangible and relatable.

Inspiring future generations is paramount to the continued flourishing of art-science fusion. By integrating art into science education and science into art curricula, we can nurture a new generation of creative thinkers who are comfortable traversing disciplinary boundaries. Interactive museum exhibits, educational workshops, and community outreach programs that blend art and science can spark curiosity and ignite a passion for learning in young minds. Imagine a school program where students cultivate bacterial colonies to create living art, exploring both microbiological principles and artistic expression simultaneously. Such initiatives foster a sense of wonder and demonstrate the interconnectedness of knowledge, encouraging students to explore diverse fields and become well-rounded individuals.

Science communication through art offers a powerful means of bridging the gap between scientific research and public understanding. Art has a unique capacity to transcend language barriers and connect with audiences on an emotional level, making complex scientific information more accessible and engaging. Visualizations, installations, and performances can translate abstract theories and data into tangible narratives, fostering a greater appreciation for the scientific process and its implications for society. Consider a multimedia installation depicting the effects of climate change on coral reefs, combining scientific data with powerful imagery and soundscapes to evoke an emotional response and inspire action. Such artistic interventions can play a crucial role in raising public awareness and driving informed decision-making on critical scientific issues.

The landscape of art-science fusion is constantly evolving, driven by technological advancements, new research discoveries, and the ever-expanding dialogue between artists and scientists. As we move forward, this interdisciplinary field holds immense potential to transform how we understand and interact with the world around us. It promises to inspire new forms of creative expression, fuel scientific innovation, and foster a deeper appreciation for the intricate beauty and complexity of the universe. By embracing the synergy between art and science, we can unlock new avenues for discovery, enhance public engagement with scientific knowledge, and cultivate a more informed and scientifically literate society. The future of art-science fusion is not just about visualizing the unseen; it's about illuminating the interconnectedness of all knowledge and inspiring us to see the world with fresh eyes. This ongoing dialogue between art and science promises a future rich with discovery, understanding, and creative expression.

6.1 Interdisciplinary Collaboration

True interdisciplinary collaboration transcends mere inspiration swapping. It necessitates a deep, ongoing dialogue and a willingness to challenge the conventional boundaries of both art and science. Think of it as a dynamic, evolving conversation where each discipline informs and transforms the other, leading to unexpected insights and novel forms of expression. This interaction isn't about scientists dictating visual representations of their work or artists superficially decorating scientific concepts. It's about a shared journey of discovery, where the scientific process itself becomes interwoven with the artistic process. This requires mutual respect, a shared vocabulary, and a commitment to navigating the inherent complexities of merging distinct fields.

For artists venturing into this realm, building a genuine understanding of the science is paramount. This goes beyond simply reading popular science articles. Engage directly with scientists, attend lectures, delve into research papers, and actively participate in discussions. Ask questions, challenge assumptions, and seek to grasp the nuances of the scientific principles that pique your interest. This deep dive into the scientific domain will provide the conceptual scaffolding needed to create truly informed and meaningful artworks.

Scientists, in turn, must be open to the transformative potential of artistic interpretation. Embrace the ambiguity and subjectivity inherent in art. Recognize that artistic expression can illuminate scientific concepts in ways that traditional scientific communication cannot. By allowing artists to explore and interpret their research, scientists can gain fresh perspectives, discover new connections, and communicate their findings to wider audiences in compelling and accessible ways. This requires relinquishing some control and embracing the possibility of unexpected outcomes.

The collaborative process itself takes many forms. Artists might embed themselves in research labs, participating in experiments and engaging in ongoing dialogue with scientists. Scientists might provide artists with access to data, microscopic imagery, or specialized equipment, offering technical guidance and conceptual insights. Joint workshops, exhibitions, and public presentations can serve as platforms for sharing ideas, fostering dialogue, and showcasing the fruits of these collaborative endeavors. The format of the collaboration should be tailored to the specific project and the individual personalities involved, ensuring a synergistic and productive partnership.

Consider, for example, an artist collaborating with a microbiologist. The artist might explore the intricate patterns and textures of microbial colonies, translating these microscopic landscapes into large-scale sculptures or immersive installations. The scientist's expertise in microbial growth and behavior could inform the artistic process, while the artist's visual interpretations could offer new perspectives on the complex dynamics of microbial communities. This reciprocal exchange of knowledge and skills leads to a richer, more nuanced understanding of the subject matter for both artist and scientist.

In the realm of quantum physics, the abstract nature of the concepts lends itself particularly well to artistic interpretation. An artist might collaborate with a physicist to explore the paradoxical nature of wave-particle duality, creating a series of photographs that capture the interplay of light and shadow, suggesting both the wave-like and particle-like properties of photons. The physicist's insights into quantum phenomena could inform the artist's visual choices, while the artist's work could provide a tangible and evocative representation of otherwise intangible concepts.

Collaboration doesn't necessitate physical proximity. Digital communication tools facilitate remote partnerships, enabling artists and scientists from different corners of the globe to connect and collaborate on projects. Sharing data, images, and ideas online allows for asynchronous communication, fostering flexibility and expanding the possibilities for interdisciplinary exchange. This global connectivity enriches the collaborative process by bringing diverse perspectives and expertise to the table.

Successful interdisciplinary collaboration requires clear communication, mutual respect, and a shared understanding of the project's goals. Establishing clear expectations from the outset is crucial. Define roles and responsibilities, agree on timelines and deliverables, and establish a framework for decision-making. Regular communication, both formal and informal, is essential for maintaining momentum and ensuring that the project stays on track. This open and transparent communication fosters trust and allows for a dynamic and productive working relationship. By embracing the challenges and rewards of interdisciplinary collaboration, artists and scientists can unlock new creative potentials and generate truly groundbreaking work that transcends the limitations of either discipline alone. It requires patience, a willingness to experiment, and a genuine desire to learn from one another. The ultimate goal is a fusion of artistic vision and scientific understanding, leading to a deeper appreciation of the interconnectedness of knowledge and the beauty of our universe.

6.2 New Media for Art & Science

The convergence of art and science has always pushed the boundaries of human expression and understanding. Historically, artists have utilized the tools and technologies available to them, from pigments derived from natural minerals to the camera obscura, a precursor to modern photography. Now, a new era of artistic exploration is unfolding, fueled by cutting-edge technologies born from scientific advancements. Digital fabrication tools like 3D printers, laser cutters, and CNC milling machines allow artists to sculpt intricate forms inspired by microscopic structures or quantum phenomena with unprecedented precision. These tools empower artists to translate complex scientific data into tangible, three-dimensional artworks, bridging the gap between the abstract and the concrete. BioArt, a fascinating subfield, utilizes living organisms as an artistic medium. Artists cultivate bacteria, fungi, or even human cells in controlled environments, harnessing their growth patterns and pigmentation to create living artworks that evolve over time. This dynamic approach challenges traditional notions of artistic creation and raises profound questions about the relationship between art, science, and life itself. It represents a direct engagement with the microscopic world, transforming scientific observation into a living, breathing artistic process.

Digital art software and virtual reality (VR) technologies offer another exciting avenue for exploring the intersection of art and science. Artists can create immersive digital environments that simulate quantum realms, allowing viewers to interact with abstract concepts like entanglement and superposition in a visceral and engaging way. Imagine stepping into a VR simulation where the laws of quantum physics govern the behavior of light and matter, experiencing firsthand the strange and beautiful world revealed by scientific inquiry. This interactive approach to art has the potential to revolutionize science education and public engagement with scientific concepts, making the invisible visible and the abstract tangible.

Sonification, the process of converting data into sound, offers yet another dimension to artistic exploration. Scientists and artists can collaborate to transform complex datasets from microscopic imaging or quantum experiments into audible soundscapes. This auditory representation of scientific data can reveal hidden patterns and relationships within the data, offering new insights for both scientists and artists. Imagine listening to the symphony of a living cell or the melodic interplay of subatomic particles, experiencing scientific data through a completely different sensory modality.

Data visualization techniques provide powerful tools for artists to transform raw scientific data into visually compelling narratives. Artists can create interactive installations where viewers can manipulate and explore complex datasets, uncovering hidden patterns and relationships. This approach not only makes scientific data more accessible to a wider audience but also empowers viewers to actively engage with the information, fostering a deeper understanding of the underlying scientific principles. By transforming abstract data into visual narratives, artists can communicate complex scientific concepts in a clear and engaging way.

Generative art, where algorithms and code are used to create artworks, provides a unique lens through which to explore scientific concepts. Artists can write algorithms that mimic natural processes, such as the growth of crystals or the evolution of biological systems, creating artworks that evolve and change over time. This approach not only creates visually stunning pieces but also offers a deeper understanding of the underlying scientific principles governing these processes. The artist becomes a programmer, using code as a brush to paint with the very fabric of reality.

Photography, both traditional and digital, continues to play a crucial role in bridging art and science. Advanced imaging techniques, such as electron microscopy and time-lapse photography, reveal the hidden wonders of the microscopic world. Artists can utilize these images as inspiration for their work or manipulate them digitally to create surreal and thought-provoking compositions. Photography allows us to see the unseen, capturing fleeting moments and revealing the intricate details of the world around us.

The intersection of art and science is a dynamic and rapidly evolving field. As new technologies emerge, artists will continue to find innovative ways to explore and express the wonders of the scientific world. This ongoing dialogue between art and science enriches both disciplines, fostering creativity, innovation, and a deeper understanding of the universe we inhabit. The fusion of artistic vision and scientific inquiry promises a future filled with breathtaking discoveries and profound artistic expressions, expanding our understanding of the world and our place within it. These new media represent not just tools, but gateways to new forms of artistic expression and scientific understanding, weaving together the threads of human creativity and the mysteries of the universe.

6.3 Inspiring Future Generations

The convergence of art and science holds a unique power—the ability to ignite curiosity and inspire future generations. By translating complex scientific concepts into accessible and engaging visual forms, art can spark interest in STEM fields among young people. This inspiration takes many forms, from captivating murals depicting cellular landscapes to interactive installations exploring the mysteries of quantum mechanics. Think of a child mesmerized by a vibrant painting of neurons firing, sparking an early interest in neuroscience. Or perhaps a teenager, intrigued by a sculpture representing the double helix structure of DNA, leading them down a path towards a career in genetics. The possibilities are endless. Art provides a crucial bridge, connecting abstract scientific principles to tangible experiences, fostering a sense of wonder and a desire to explore.

Cultivating this connection requires proactive initiatives. Schools can integrate art-science projects into their curricula, providing students with opportunities to create their own artistic interpretations of scientific concepts. Imagine a classroom where students build models of molecules, design posters explaining the water cycle, or even compose musical pieces inspired by the rhythms of nature. These hands-on experiences not only solidify their understanding of scientific principles but also nurture their creativity and critical thinking skills. Museums and science centers can play a vital role as well, hosting interactive exhibits that blend art and science, offering workshops that teach scientific visualization techniques, and creating spaces where artists and scientists can collaborate. These institutions can become hubs for interdisciplinary exploration, inspiring young visitors to see the connections between seemingly disparate fields and to envision themselves as future innovators.

Mentorship programs pairing established artists and scientists with aspiring young creatives can further foster this burgeoning interest. Imagine a young artist shadowing a biologist, learning about the intricate structures of microorganisms and translating that knowledge into a series of stunning microscopic portraits. Or picture a budding scientist working alongside a digital artist, using coding and algorithms to create visualizations of complex datasets, transforming raw data into captivating works of art. These mentorship opportunities provide invaluable guidance, encouragement, and practical experience, helping young people develop the skills and confidence to pursue their passions.

Furthermore, leveraging the power of digital media can significantly expand the reach and impact of art-science initiatives. Online platforms can host virtual galleries showcasing student artwork, interactive tutorials explaining scientific concepts through artistic mediums, and forums where young people can connect with mentors and peers. Consider a virtual reality experience that allows students to explore the human body at a cellular level, guided by artistic representations of different organelles and their functions. Or envision an online platform where students can create and share their own digital artwork inspired by scientific discoveries, fostering a global community of young art-science enthusiasts. The digital realm offers unprecedented opportunities to connect, collaborate, and inspire, breaking down geographical barriers and empowering young people worldwide to engage with art and science in dynamic new ways.

Supporting young artists and scientists doesn't just mean providing them with resources and opportunities; it also requires cultivating a culture that values interdisciplinary thinking. We need to move beyond traditional silos and encourage collaboration between artists, scientists, educators, and policymakers. This could involve establishing grants specifically for art-science projects, incorporating art-science integration into national education standards, and creating platforms for dialogue and knowledge sharing between these often-separated communities. By fostering a collaborative ecosystem, we can create an environment where young people feel empowered to explore the intersections of art and science, leading to groundbreaking discoveries and innovative solutions to the challenges facing our world.

The future of art and science lies in the hands of the next generation. By nurturing their curiosity, providing them with the tools and resources they need to succeed, and fostering a culture that values interdisciplinary exploration, we can unlock their full potential. Investing in these young minds isn't just about advancing art and science; it's about investing in a brighter future for all. It's about inspiring a generation of creative thinkers, problem solvers, and innovators who can bridge the gap between disciplines, address complex global challenges, and shape a world where art and science work together to illuminate the wonders of our universe.

6.4 Science Communication Through Art

Effective science communication hinges on the ability to translate complex information into accessible and engaging formats. Art offers a powerful avenue for achieving this, transforming abstract concepts into tangible experiences. Consider the intricate dance of subatomic particles, a realm governed by the often counterintuitive laws of quantum mechanics. Visualizing such phenomena through artistic mediums—sculpture, painting, digital art—can bridge the gap between scientific understanding and public perception. An artist might represent the probabilistic nature of quantum states through a series of overlapping, translucent forms, creating a visual echo of superposition. Or perhaps the entanglement of particles could be depicted by interconnected lines and shapes, subtly shifting and responding to one another across the canvas, mirroring the particles' intertwined fates. Such artistic interpretations not only offer aesthetic experiences but also serve as powerful tools for conveying complex scientific principles.

The intersection of art and science allows for a deeper exploration of scientific concepts, particularly those dealing with the invisible or abstract. Microscopic imaging techniques have unveiled a hidden universe teeming with intricate structures and vibrant life. Electron micrographs reveal the delicate architecture of diatoms, the complex machinery of cells, and the surprisingly beautiful symmetry of viruses. These images, often breathtaking in their complexity and detail, serve as a natural springboard for artistic expression. Sculptors might create three-dimensional representations of these microscopic forms, magnifying them to scales that allow viewers to appreciate their hidden elegance. Painters can translate the vibrant colors and textures revealed by microscopy into captivating artworks, offering a glimpse into the world beyond our normal perception. These artistic renditions can ignite curiosity and spark a deeper interest in the science behind the imagery.

Beyond static representations, artistic mediums can also capture the dynamic processes that drive the natural world. Time-lapse photography and animation, paired with scientific data, can visually depict the intricate unfolding of biological phenomena, from cell division to the growth of crystals. Such dynamic art forms can convey the constant flux and transformation that characterizes life and the universe itself, enriching our understanding of these complex systems. Imagine a time-lapse video showing the growth of a fungal network, its thread-like hyphae reaching out and intertwining, accompanied by a soundtrack that translates the electrical signals passing through the network. This kind of multi-sensory experience offers a richer, more engaging exploration of scientific processes.

The power of art to communicate science extends beyond visual representation. It can also encompass other sensory modalities, creating immersive experiences that deepen understanding and engagement. Sound art installations, for example, can translate scientific data into auditory experiences. The rhythmic firing of neurons, the complex oscillations of climate patterns, or the subtle vibrations of molecular structures can be transformed into captivating soundscapes, offering a unique perspective on scientific data. This allows individuals who may not easily grasp visual representations to access and connect with complex scientific concepts through a different sensory pathway. Moreover, tactile sculptures can represent complex protein structures, allowing viewers to physically interact with and explore the intricacies of molecular biology.

By engaging multiple senses, art can create a more holistic and memorable learning experience. This multi-sensory approach is particularly effective in educational settings, making science more accessible to diverse learners. Interactive installations in museums and science centers, incorporating visual, auditory, and tactile elements, can transform passive observation into active exploration. Visitors can engage with scientific concepts in a more visceral and intuitive way, fostering a deeper understanding and appreciation for the scientific world. Imagine an exhibit on the human microbiome that combines microscopic images of bacteria with a soundscape of their metabolic activity and a tactile model of the human intestinal lining. This kind of multi-faceted experience allows visitors to explore the microbiome from multiple perspectives, creating a richer and more memorable learning experience.

Ultimately, the fusion of art and science serves as a powerful tool for fostering dialogue and promoting scientific literacy. Art provides a common language, accessible to audiences with varying levels of scientific background. It can spark curiosity, stimulate critical thinking, and inspire a deeper engagement with scientific issues. Through artistic interpretations, complex scientific concepts become relatable, humanized, and integrated into the broader cultural landscape. This fosters a more informed and engaged public, better equipped to understand and appreciate the scientific discoveries that shape our world.

6.5 The Evolving Landscape

The convergence of art and science is not a static endpoint, but a dynamic, ever-shifting landscape. This continuous evolution is fueled by advancements in both fields, opening up new avenues for creative exploration and pushing the boundaries of how we perceive and interpret the world around us. Consider the rapid development of imaging technologies, from advanced microscopy techniques like cryo-electron microscopy, capable of visualizing individual molecules, to the breathtaking imagery captured by the James Webb Space Telescope, unveiling the universe's deepest secrets. These technological leaps provide artists with fresh perspectives and a wealth of visual data to inspire their work, leading to novel artistic interpretations of scientific discoveries.

This evolving landscape is also characterized by an increasing blurring of the lines between artist and scientist. We are witnessing a rise in "art-science hybrids"—individuals trained in both disciplines who seamlessly integrate scientific principles into their artistic practice. These individuals possess a unique ability to bridge the gap between two seemingly disparate worlds, creating works that are both aesthetically compelling and scientifically informed. Think of artists who cultivate living organisms as part of their artistic medium, or those who use algorithms and data visualization to generate complex and evocative art pieces. These art forms not only showcase the beauty inherent in scientific processes, but also invite us to contemplate the ethical and philosophical implications of these advancements.

The rise of digital technologies and the internet has further democratized access to scientific information and artistic tools. Open-source software, online tutorials, and digital fabrication techniques have empowered a new generation of artists and scientists to collaborate and share their work with a global audience. This increased accessibility fosters a vibrant exchange of ideas, leading to more diverse and innovative artistic expressions of scientific concepts. Imagine an online platform where artists and scientists can collaborate on projects, sharing data sets, visualizations, and artistic interpretations. Such platforms have the potential to accelerate the pace of discovery and spark new forms of creative expression.

Furthermore, the evolving landscape is marked by a growing recognition of the importance of art in science communication. As scientific research becomes increasingly specialized and complex, the need for effective communication with the public becomes even more crucial. Art offers a powerful tool for translating complex scientific concepts into accessible and engaging narratives. Visual art, in particular, can transcend language barriers and evoke emotional responses, making scientific discoveries more relatable and memorable for broader audiences. Consider the impact of scientific visualizations, which can transform abstract data into stunning visual displays, or the role of storytelling in conveying the human element behind scientific research. These approaches not only educate but also inspire curiosity and wonder about the natural world. Looking ahead, we can anticipate even more profound integration of art and science. The development of new technologies, like virtual and augmented reality, offers exciting possibilities for immersive artistic experiences that seamlessly blend scientific data with artistic expression. Imagine exploring the intricate structure of a cell through a virtual reality experience, or interacting with a quantum phenomenon through an augmented reality installation. These technologies have the potential to revolutionize how we engage with scientific concepts, making them more accessible, interactive, and personally meaningful.

Finally, this evolving landscape requires a shift in how we educate and train future generations. Interdisciplinary programs that encourage collaboration between artists and scientists are essential for fostering a new breed of creative thinkers who can effectively navigate the complexities of the 21st century. By providing students with opportunities to explore the intersection of art and science, we can empower them to become innovators, communicators, and problem-solvers who can address the grand challenges facing our society. This includes promoting STEAM education (Science, Technology, Engineering, Arts, and Mathematics) that integrates artistic thinking and creative problem-solving into traditional STEM curricula. This holistic approach not only equips students with the technical skills they need, but also cultivates the creative mindset essential for innovation and discovery. The future of art and science is not merely a convergence, but a vibrant and dynamic interweaving, constantly evolving and pushing the boundaries of human understanding.

7 Conclusion: Art Illuminating Science

The journey through the microscopic landscapes and the enigmatic quantum realms has revealed a profound connection between art and science. We've witnessed how artists, inspired by scientific inquiry, translate complex concepts into visceral experiences. The intricate structures of cells, once hidden from the naked eye, have blossomed into vibrant artistic expressions. The counterintuitive principles of quantum physics, often shrouded in mathematical abstraction, have materialized into tangible forms through the artist's lens. This transformative power of art underscores its crucial role, not merely as a reflection of scientific discovery, but as an active participant in its exploration.

Consider the delicate dance between a scientist peering through a microscope and an artist meticulously crafting a sculpture. The scientist observes, analyzes, and documents the intricate details of a cellular structure. The artist, drawing inspiration from these observations, translates the scientific data into a three-dimensional form, emphasizing texture, shape, and the interplay of light and shadow. This artistic interpretation doesn't simply replicate the scientific image; it imbues it with a new dimension of meaning, making it accessible and engaging for a wider audience. The artwork becomes a conduit for understanding, bridging the gap between the scientific realm and the human experience.

The abstract nature of quantum physics presents a unique challenge for artistic representation. Concepts like entanglement and superposition defy easy visualization. Yet, artists have risen to this challenge, employing abstraction, symbolism, and innovative techniques to capture the essence of these phenomena. A swirling vortex of color might represent the probabilistic nature of quantum states. Interconnected lines and forms might symbolize the interconnectedness of entangled particles. These artistic interpretations, while not literal depictions, offer a powerful means of grappling with complex scientific ideas, prompting reflection and stimulating new ways of thinking about the universe.

Moreover, the fusion of art and science fosters a deeper appreciation for both disciplines. By engaging with scientific concepts through an artistic lens, we develop a more nuanced understanding of the scientific principles at play. The artistic process encourages us to look beyond the data and equations, to connect with the underlying beauty and wonder of the natural world. Conversely, scientific inquiry can inspire new avenues of artistic exploration, pushing the boundaries of creative expression. The precision and meticulousness inherent in scientific investigation can inform artistic practice, while the imaginative and intuitive nature of art can spark new scientific questions and hypotheses.

The convergence of art and science extends beyond the creation of individual artworks. It has profound implications for science communication and education. Artistic interpretations can make complex scientific concepts accessible to a wider audience, fostering greater scientific literacy and engagement. Visualizations of microscopic life or quantum phenomena can capture the imagination in ways that traditional scientific communication often struggles to achieve. This accessibility is crucial for bridging the gap between scientific research and public understanding, promoting informed decision-making and fostering a greater appreciation for the role of science in society.

Ultimately, the interplay between art and science illuminates the interconnectedness of human knowledge and experience. Art provides a powerful language for expressing and interpreting scientific discoveries, expanding our perspectives and enriching our understanding of the world around us. It encourages us to embrace curiosity, to question assumptions, and to explore the universe with both intellectual rigor and creative wonder. As we continue to unravel the mysteries of the cosmos, the partnership between art and science will undoubtedly play an increasingly vital role in shaping our understanding of our place within it. The dialogue between these two seemingly disparate disciplines reminds us that knowledge is not confined to the laboratory or the studio, but emerges from the dynamic interplay of observation, imagination, and expression. This synergistic relationship, where art illuminates science and science inspires art, holds immense potential for unlocking new discoveries and fostering a deeper appreciation for the beauty and complexity of the universe. It is in this fertile ground of interdisciplinary exploration that we can truly grasp the profound interconnectedness of all knowledge and unlock new pathways for understanding ourselves and the world we inhabit.

7.1 The Power of Visual Language

Visual language, in its broadest sense, encompasses the ways we communicate through visual elements. Think about the cave paintings of Lascaux—early humans, driven by a need to record and share their world, crafted images that spoke volumes about their lives, beliefs, and understanding of their environment. This primal form of storytelling, using pigments and surfaces to convey meaning, laid the foundation for the sophisticated visual language we engage with today. In the context of this book, exploring the microscopic and quantum realms, visual language becomes the crucial bridge between complex scientific concepts and our capacity to grasp them. It's the vehicle through which the invisible becomes visible, the abstract tangible.

Consider the intricate structure of a diatom, a single-celled algae revealed under the microscope. Its glassy shell, etched with delicate patterns, is a testament to the elegance of natural engineering. A scientist might describe its structure through precise measurements and biological terminology, but an artist can translate these details into a visual experience—a finely detailed drawing, a shimmering glass sculpture, a photograph capturing the play of light on its crystalline surface. This artistic interpretation doesn't replace scientific understanding; it enhances it, adding layers of emotional resonance and aesthetic appreciation. The diatom becomes more than a specimen; it transforms into a source of wonder and inspiration.

Now shift your perspective from the microscopic to the cosmic, to the realm of quantum physics. Concepts like superposition and entanglement defy easy visualization. They exist in a world governed by probabilities and counterintuitive interactions, a world far removed from our everyday sensory experiences. Yet, artists find ways to express these complex ideas through visual metaphors. The blurring of lines in a painting might represent the uncertainty principle, the overlapping of colors the superposition of quantum states. A sculptor might create an interconnected installation to symbolize the entanglement of particles, their fates inextricably linked across vast distances. These artistic explorations provide a tangible entry point into a realm that might otherwise remain inaccessible to the non-scientist.

Photography, in particular, plays a vital role in bridging this gap between the seen and unseen. Macro and micro-photography allow us to witness the intricate details of the natural world, from the compound eye of an insect to the crystalline structure of a mineral. Astrophotography captures the grandeur of distant galaxies, nebulae painted across the cosmic canvas. These photographic images serve not only as scientific records but also as aesthetic objects, sparking curiosity and inspiring a sense of awe. They remind us of the beauty and complexity that exist beyond the limits of our unaided perception.

Moreover, digital tools have expanded the possibilities of visual language. Software allows artists to manipulate images, creating visualizations of abstract scientific data, simulations of dynamic processes, and immersive virtual environments. These digital creations can convey information in ways that traditional media cannot, offering interactive explorations of scientific concepts and allowing viewers to engage with complex data in a more intuitive and personalized way. Imagine exploring the inner workings of a cell through a virtual reality experience, or visualizing the flow of quantum fields through an animated artwork. These digital tools empower artists to become translators of scientific knowledge, transforming data into compelling visual narratives.

The power of visual language lies not merely in its ability to represent, but also in its capacity to evoke. A photograph of a melting glacier can be more impactful than a graph charting temperature increases. A sculpture depicting the interconnectedness of life can resonate more deeply than a scientific paper describing ecological relationships. By tapping into our emotions and aesthetic sensibilities, visual language can bypass the intellectual barriers that often surround complex scientific concepts. It can inspire wonder, curiosity, and a deeper appreciation for the world around us, from the smallest microbe to the largest galaxy. It is through this emotional connection that scientific understanding truly takes root and flourishes. This fusion of art and science, facilitated by the power of visual language, is not simply about making science more accessible; it's about enriching our understanding of the world in its entirety, fostering a deeper connection between ourselves and the universe we inhabit.

7.2 Expanding Perspectives

True understanding often requires a shift in perspective, a willingness to step outside familiar frameworks and embrace new ways of seeing. Consider the limitations of our everyday perception. We are bound by the scales of our senses, unable to directly perceive the intricate dance of molecules or the vastness of cosmic expansion. Art, however, offers a unique bridge, translating complex scientific concepts into tangible, evocative forms. It allows us to engage with the unseen world, not through data points and equations, but through color, texture, and form. This transformative power of art extends beyond simply illustrating scientific principles; it opens up new avenues for inquiry, sparking curiosity and prompting us to question our assumptions about the nature of reality.

By exploring scientific concepts through an artistic lens, we gain a deeper appreciation for the interconnectedness of knowledge. The intricate patterns revealed by a microscope can inspire a textile artist, translating cellular structures into woven masterpieces. The ethereal dance of subatomic particles can find expression in a sculptor's hands, transforming abstract quantum principles into tangible forms. This cross-pollination of ideas enriches both disciplines. Scientists may find new ways to visualize their data, while artists discover fresh sources of inspiration and innovative techniques. The fusion of scientific understanding with artistic expression generates a powerful synergy, pushing the boundaries of both fields and opening new vistas of understanding.

The ability of art to evoke emotion plays a crucial role in expanding perspectives. While scientific data provides objective information, art taps into our subjective experience, fostering a visceral connection with complex concepts. A painting depicting the vibrant hues of a coral reef can inspire a sense of wonder and a deeper appreciation for the delicate balance of marine ecosystems. A sculpture representing the dynamic forces of plate tectonics can evoke a sense of awe at the Earth's powerful geological processes. This emotional engagement is essential for fostering genuine understanding and inspiring action. It's not enough to simply know the science; we must feel it, internalize it, and allow it to shape our perceptions of the world.

Expanding perspectives also involves challenging conventional modes of representation. Traditional scientific visualizations, while informative, can sometimes feel sterile and detached. Art, on the other hand, offers a limitless palette of expressive possibilities. Photography can capture the ephemeral beauty of scientific phenomena, freezing moments in time and revealing hidden details. Digital manipulation allows artists to create surreal landscapes inspired by microscopic imagery, transforming scientific data into otherworldly visions. Sculpture can give form to invisible forces, allowing us to interact with abstract concepts in a tangible way. By embracing these diverse forms of artistic expression, we can move beyond the limitations of traditional scientific representation and engage with complex ideas in a more meaningful and impactful way.

Furthermore, the collaborative nature of art-science projects fosters interdisciplinary dialogue and expands perspectives within the scientific community itself. By working alongside artists, scientists are encouraged to think creatively about their research, to consider alternative ways of visualizing their data, and to communicate their findings to a wider audience. This interaction can lead to new insights and discoveries, as scientists gain fresh perspectives on their work through the eyes of artists. The process of translating scientific concepts into artistic forms can reveal hidden patterns, suggest new lines of inquiry, and inspire innovative approaches to scientific investigation.

Expanding perspectives also entails recognizing the inherent limitations of our knowledge. Science is a constantly evolving process, with new discoveries continually reshaping our understanding of the universe. Art can help us grapple with this uncertainty, embracing the unknown and fostering a sense of wonder about the mysteries that remain. Abstract art, for instance, can provide a visual language for exploring the complexities of quantum mechanics, where uncertainty and probability reign supreme. By acknowledging the limits of our current understanding, we open ourselves to new possibilities and cultivate a spirit of continuous inquiry.

Finally, expanding perspectives means acknowledging the diverse ways in which individuals engage with scientific concepts. Not everyone learns through textbooks and equations. Art offers an alternative pathway to understanding, providing a visual and emotional entry point for those who may feel intimidated by traditional scientific discourse. By embracing diverse forms of communication and representation, we can make scientific knowledge more accessible and inclusive, fostering a greater appreciation for science across all segments of society. The marriage of art and science offers a powerful tool for broadening participation in scientific discourse and empowering individuals to engage with complex ideas in meaningful ways. This inclusive approach to science communication is crucial for fostering informed decision-making and addressing the complex scientific challenges facing our world today.

www.ingramcontent.com/pod-product-compliance
Ingram Content Group UK Ltd.
Pitfield, Milton Keynes, MK11 3LW, UK
UKHW021436240125
4283UKWH00041B/630

9 798348 415624